# THE PURSUIT OF PERFECT PACKING

## SECOND EDITION

# THE PURSUIT OF PERFECT PACKING

## SECOND EDITION

## TOMASO ASTE

*AUSTRALIAN NATIONAL UNIVERSITY*
*CANBERRA, AUSTRALIA*

## DENIS WEAIRE

*TRINITY COLLEGE*
*DUBLIN, IRELAND*

Taylor & Francis
Taylor & Francis Group
New York   London

Taylor & Francis is an imprint of the
Taylor & Francis Group, an **informa** business

CRC Press
Taylor & Francis Group
6000 Broken Sound Parkway NW, Suite 300
Boca Raton, FL 33487-2742

International Standard Book Number-13: 978-1-4200-6817-7 (Hardcover)

---

**Library of Congress Cataloging-in-Publication Data**

---

Aste, Tomaso.
    The pursuit of perfect packing / Tomaso Aste and Denis Weaire. -- 2nd ed.
      p. cm.
    Includes bibliographical references and index.
    ISBN 978-1-4200-6817-7 (alk. paper)
    1. Combinatorial packing and covering. I. Weaire, D. L. II. Title.

QA166.7.A78 2008
511'.6--dc22
                                           2007044016

---

**Visit the Taylor & Francis Web site at**
**http://www.taylorandfrancis.com**

**and the CRC Press Web site at**
**http://www.crcpress.com**

*Dedicated to Colette and Tiziana*

# Contents

# Preface to First Edition

There are many things which might be packed into a book about packing. Our choice has been eclectic. Around the mathematical core of the subject we have gathered examples from far and wide.

It was difficult to decide how to handle references. This is not intended as a heavyweight monograph or an all-inclusive handbook, but the reader may well wish to check or pursue particular topics. We have tried to give a broad range of general references to authoritative books and review articles. In addition we have identified the original source of many of the key results which are discussed, together with enough clues in the text to enable other points to be followed up, for example with a biographical dictionary.

Thanks are due to many colleagues who have helped us, including Nicolas Rivier (a constant source of stimulation and esoteric knowledge), Stefan Hutzler and Robert Phelan. Rob Kusner and Jörg Wills made several suggestions for the text, which we have adopted.

Denis Weaire benefited from the research support of Enterprise Ireland and Shell during the period in which this was written, and Tomaso Aste was a Marie Curie Research Fellow of the European Union (EU) during part of it.

**Tomaso Aste**
**Denis Weaire**

# Preface to Second Edition

Six years after our first Pursuit, we decided to offer a revised and augmented edition. Some of the stories that we told have continuations and sequels worth the telling. It is also a fine opportunity to weed out mistakes and misconceptions, here and there. Moreover, one of the structures closest to our hearts now looms large on the Beijing Olympic townscape, surely a story to be added.

We are grateful to the publishers for this opportunity, and to many colleagues for generous help. They include Tristram Carfae (Arup), Chris Bosse, Renaud Delannay, Tiziana Di Matteo, Nick Cook, Bruce Hunt, John Nye. The Australian National University, Canberra is also to be thanked for hospitality to Denis Weaire (DW) during this process of revision.

**Tomaso Aste**
**Denis Weaire**

*On poetry and geometric truth,*
*And their high privilege of lasting life,*
*From all internal injury exempt,*
*I mused; upon these chiefly: and at length,*
*My senses yielding to the sultry air,*
*Sleep seized me, and I passed into a dream.*

William Wordsworth
The Prelude

# Chapter 1

# How Many Sweets in the Jar?

The half-empty suitcase or refrigerator is a rare phenomenon. We seem to spend much of our lives squeezing things into tight spaces, and scratching our heads when we fail. The poet might have said: *Packing and stacking we lay waste our days.*

To the designer of circuit boards or software the challenge carries a stimulating commercial reward: savings can be made by packing things well. How can we best go about it, and how do we know when the optimal solution has been found?

This has long been a teasing problem for the mathematical fraternity, one in which their delicate webs of formal argument somehow fail to capture much certain knowledge. Their frustration is not shared by the computer scientist, whose more rough-and-ready tactics have found many practical results.

Physicists also take an interest, being concerned with how things fit together in nature. And many biologists have not been able to resist the temptation to look for a geometrical story to account for the complexities of life itself. So our account of packing problems will range from atoms to honeycombs in search of inspiration and applications.

Unless engaged in smuggling, we are likely to pack our suitcase with miscellaneous items of varying shape and size. This compounds the problem of how to arrange them. The mathematician would prefer to consider identical objects, and an infinite suitcase. How then can oranges be packed most tightly, if we do not have to worry about the container? This is a celebrated question, associated with the name of one of the greatest figures in the history of science, Johannes Kepler, and highlighted by David Hilbert at the start of the twentieth century. It is still topical today.

**Figure 1.1**   Stacking casks of Guinness.

### Hilbert's 18th problem

In 1900, David Hilbert presented to the International Mathematical Congress in Paris a list of 23 problems which he hoped would guide mathematical research in the twentieth century. The 18th problem was concerned with sphere packing and space-filling polyhedra.

I point out the following question (...) important to number theory and perhaps sometimes useful to physics and chemistry: *How one can arrange most densely in space an infinite number of equal solids of given form,* e.g., spheres with given radii or regular tetrahedra with given edges (or in prescribed positions), *that is, how can one so fit them together that the ratio of the filled to the unfilled space may be as great as possible?*

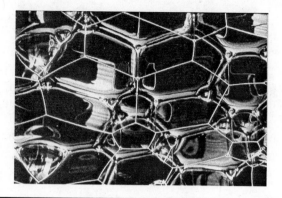

**Figure 1.2** Soap bubbles.

More subtle goals than that of maximum density may be invoked. When bubbles are packed tightly to form a foam, as in a glass of beer, they can adjust their shapes and they do so to minimize their surface area. So in this case, the volume of each bubble is fixed and it is the total area of the interfaces between the bubbles that is minimized. This is a packing of bubbles, but also a partitioning of space into compartments or cells.

The history of ideas about packing is peopled by many eminent and colorful characters. An English reverend gentlemen is remembered for his

**Figure 1.3** Packing on a grand scale. (Wright, T., 1750, *The Cosmos*. With permission.)

experiments in squashing peas together in the pursuit of geometrical insights. A blind Belgian scientist performed by proxy the experiments that laid the ground rules for serious play with bubbles. An Irishman of unrivalled reputation for dalliance (at least among crystallographers) gave us the rules for the random packing of balls. A Scotsman who was the grand old man of Victorian science was briefly obsessed with the parsimonious partitioning of space.

All of them shared the curiosity of the child at the church bazaar: *How many sweets are there in the jar?*

# Chapter 2

---

# Loose Change and Tight Packing

---

## 2.1 A Teasing But Tractable Problem

The pain and pleasure of intractable centuries-old problems will permeate much of this book, but let us start with an easier one. It will serve to introduce the style of thinking that is involved and the ingredients out of which a mathematician might bake his pie. But is not trivial: the proof is given in Appendix A. The reader may wish to read on without looking at it, and try to dream up a demonstration, before looking at the one provided.

## 2.2 A Handful of Coins

An ample handful of loose change, spread out on a table, will help us understand some of the principles of packing. This will serve to introduce some of the basic notations of the subject, before we tackle the complexity of three dimensions and the obscurities of higher dimensions.

Let us discard the odd-shaped coins which are becoming fashionable; we want hard circular discs. Coins come in various sizes, so let us further simplify the problem by selecting a set of equal size. About 10 will do. The question is: *How could a large number of these coins be arranged most tightly?*

If we do it in three-dimensional space, then obviously the well known bank-roll is best for any number of coins. But here we restrict ourselves to two-dimensional packing on the flat surface of the table.

Three coins fit neatly together in a triangle, as in Figure 2.1a. There is no difficulty in continuing this strategy, with each coin eventually contacting

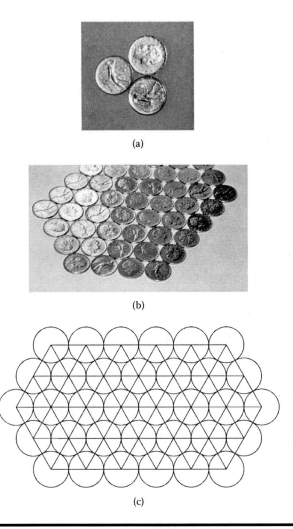

(a)

(b)

(c)

**Figure 2.1** Three equal discs fit tightly in an equilateral triangle (a). This configuration can be extended to generate the triangular close packing (b). This is the
densest possible arrangement of equal discs, having packing fraction $\rho = 0.9068\ldots$.
The regular pattern drawn by joining the centers of touching discs is the triangular
lattice (c).

six neighbors. A pleasant pattern soon emerges: the triangular close packing
in two dimensions.

The fraction of table covered by coins is called the packing fraction,
and in this case it is

$$\rho = \frac{\text{Area covered}}{\text{Total area}} = \frac{\pi}{\sqrt{12}} = 0.9068\ldots \tag{2.1}$$

This must surely be the largest possible value. But can we prove it? That is often where the trouble starts, but not in this case. A proof can be constructed as shown in Appendix A. It illustrates some of the mathematical ingredients that may, in favorable cases, furnish a proof of what is best.

In his book, which covers many such problems,[1] Fejes Tóth attributes the first proof of this result to the Scandinavian mathematician Thue, who in 1892 discussed the problem at the Scandinavian National Science Congress and in 1910 published a longer proof. C. A. Rogers expresses some objections to these proofs: "It is no easy task to establish certain compactness results which he takes for granted."[2]

What we present in Appendix A is based on one of the proofs given by Fejes Tóth.

Here we have a pattern which, were we not to construct it in the imagination, presents itself to us in daily life, in the packing and stacking of drinking straws and other cylindrical objects, or the layout of eggs in a carton. It would even form spontaneously, at least locally, if we shuffled our coins together for long enough (Figure 2.2).

There is a research group in Northern France which does just this sort of experiment, using a gigantic air table. This is a surface with small holes through which air is pumped, so that flat objects can glide upon it without friction (Figure 2.3).

Later we will encounter other problems which look very similar to this one, but few of them submit to such a straightforward argument. Indeed a false sense of security may be induced by this example. It turns out that almost any variation of the problem renders it more mysterious.

So what is so special about the packing of these discs? It is the convenient fact that there is a best *local* packing which can be extended without variation to the whole structure. If we restore our handful of *mixed* change, or try to stack oranges in three dimensions instead, no such elementary argument will work.

The problem becomes one of a *global* optimization, which cannot always be achieved by *local* optimization. It takes its place alongside many others which practical people have to face. Under specified conditions how do we maximize some global quantity? Given life and liberty how do we best pursue happiness, allowing for the limitations of human nature and the competing demands of individuals?

As every democrat knows, global optimization is a matter of difficult compromise. Perfection is rarely achieved, and how do you know when you have it?

---

[1] Fejes, Tóth L., 1953, *Lagerungen in der Ebene auf der Kugel und im Raum (Die Grundlehren der Math. Wiss. 65)* (Berlin: Springer).

[2] Rogers, C. A., 1964, *Packing and Covering* (Cambridge: Cambridge University Press).

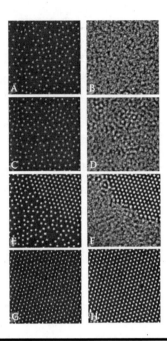

**Figure 2.2** Spontaneous clustering into the triangular lattice for bearings on a plane shaken vertically from high (top) to low accelerations (bottom). Single images (right column) and second averaged images (left column). (Courtesy of J. S. Olafsen and J. S. Urbach.)

Theorists, when confronted by this kind of conflict of local and global requirements often call it *frustration*, aptly enough.[3] Our subject is full of frustration.

## 2.3 Order and Disorder

We have placed our coins in a highly ordered structure. This seems natural for identical objects; prejudice runs strongly in favor of order, in the search for an optimal packing.

Here we mean that kind of order which is called *crystalline* and is the preoccupation of an entire profession, as we shall recall in Chapter 13. A crystal has a regularly repeating structure: if you have seen one bit, you've seen it all.

The Nobel Prizewinning physicist Philip W. Anderson raised the question in his book *Concepts of Solids*: Why crystals? Why order? A satisfactory answer has not been given to this elusive question, and the subsequent discovery of *quasicrystals* (Chapter 13) has further complicated this issue.

---

[3] Sadoc, J.-F. and Mosseri, R., 1997, *Frustration Géométrique* (Paris: Éditions Eyrolles).

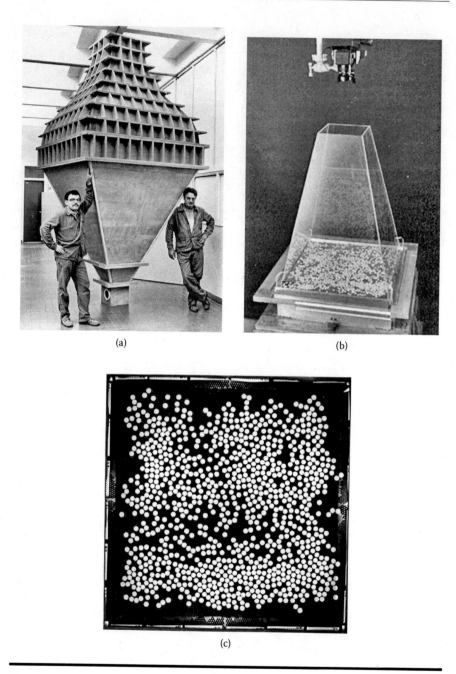

(a)

(b)

(c)

**Figure 2.3** (a,b) The giant air table machine (4.5 m high) constructed in Rennes by D. Bideau and others, in order to investigate two-dimensional packings. The two men in the photograph are the builders of the system. (c) An example of disk packing generated by this machine. (Courtesy of the Granular Group, GMCM, UMR-CNRS 6626, Université Rennes 1.)

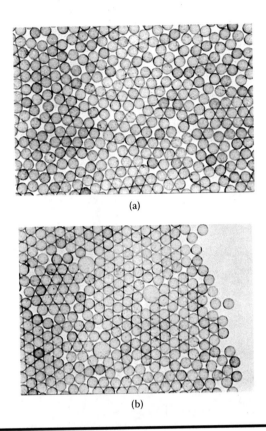

(a)

(b)

**Figure 2.4**   A disordered assembly of equal discs (a), and a packing of discs with two different sizes (b).

Disordered packings do occur in practice even for identical objects and they are not just a defective version of an ordered ideal (Figure 2.4). We will see that they have their own character, and pose fresh problems (Chapter 5). They may sometimes be called "random," but in this subject the word does not have any strict technical meaning.

Our identical coins, when drawn together, order spontaneously. But if a mixture of sizes or shapes is used, disorder persists. A more typical handful of change will suffice for us to see this. In principle, some meticulous rearrangement might create a better ordered structure, but human intervention would be required to compose it.

So in this book we are concerned not just with ideal order, but also with real disorder, wherever it takes an interesting, reproducible form.

But for the moment let us be idealistic egg-heads, and confront a problem of ideal packing that is both historical and topical: What is the best packing of equal spheres?

# Chapter 3

# Hard Problems with Hard Spheres

## 3.1  The Greengrocer's Dilemma

Now we will use our coins to buy a heap of oranges or a bag of ball bearings, and explore their packing. It is much more difficult to visualize the possibilities that they present, in the mind's eye or in reality. But one thing becomes quite clear at an early stage: no amount of shaking the bag will cause the balls to come together in an elegant ordered structure. By the same token, the greengrocer must take time and care to stack his oranges neatly (Figure 3.1). Is his stacking the densest possible? Of course, this is not his objective. For our greengrocer, considerations of stability and aesthetics are paramount, in the pursuit of profit. But if he is an amateur mathematician perhaps he might just wonder. . . .

## 3.2  Ordered Close Packing—The Kepler Problem

In the greengrocer's shop, the proprietor, unconcerned about theory, has stacked his oranges in a neat pyramidal pile. He proceeds by first laying out the fruit in the manner of the coins of Chapter 2, and observes that this creates resting-places for the next layer, and so on (Figure 3.2). The packing fraction is $\pi/\sqrt{18} = 0.740\,48\ldots$. Is this the best that can be done?

It has been remarked that *all mathematicians think and all physicists know that this is the best*. Some metallurgists have gone one step further by declaring in textbooks that such an assertion has been *proved*. At least prior to August 1998, such a statement was quite incorrect. This is the long-standing Kepler Problem.

(a)

(b)

**Figure 3.1** At the greengrocery.

Let us resort to theoretical argument, following the line of Chapter 2. In three dimensions the natural counterpart of the trio of coins in mutual contact is the regular tetrahedron (Figure 3.3). If these symmetric fourfold units could be tightly assembled we might well expect them to give us the best possible structure in three dimensions, and its packing fraction would be $\sqrt{18}[\cos^{-1}(\frac{1}{3}) - \pi/3] = 0.7797\ldots$. But they cannot be so combined. This strategy fails.

Such a packing would necessarily involve tetrahedra which share a common edge, but the angle between two faces (dihedral angle, $\theta = \cos^{-1}(\frac{1}{3})$) does not allow this, as Figure 3.3 shows. This is a rather straightforward case of *frustration* at work: four spheres are best placed in symmetric tetrahedral positions, but it cannot be so for a great many (Figure 3.3).

**Figure 3.2** This is the sphere packing commonly found on fruit stands, in piles of cannon balls on war memorials and in the crystalline structures of many materials.

An excellent strategy to relieve this frustration would be to relax the condition that the tetrahedra be regular (perfectly symmetric), as they must be if all four balls are in contact with each other. We shall return to this possibility in Chapter 13. Another strategy of the idealist is to curve space in such a way as to allow tetrahedra to fit perfectly. This sort of game seems artificial to most of us, since we cannot readily curve the space around us (stars and galaxies can do it). But for two-dimensional packings it is quite natural. For example, we may choose to perform a packing on the surface of a sphere (Chapter 18).

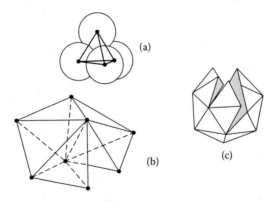

**Figure 3.3** Four spheres can be assembled mutually in contact on the vertices of a regular tetrahedron (a). This disposition is the closest possible, with packing fraction $\rho = 0.7797\ldots$. But regular tetrahedra cannot be combined to fill the whole space. The angle between the two faces of the tetrahedron is $\theta = \cos^{-1}(\frac{1}{3}) = 2\pi/5.104\ldots$, which means that around a common edge one can dispose five tetrahedra, but an interstice of $2\pi - 5\cos^{-1}(\frac{1}{3}) = 0.128\ldots$ will remain (b). Twenty contacting tetrahedra can be disposed around a common vertex (c).

**Figure 3.4** Thomas Harriot (1560–1621). (Courtesy of Trinity College, Oxford.)

## 3.3 The Kepler Conjecture

Science is notoriously dependent on military motives. In 1591, Walter Raleigh needed a formula to count the number of cannonballs in a stack. His friend Thomas Harriot,[1] who was his surveyor/geographer on the second expedition to Virginia, developed a formula without much difficulty and made a study of close packing (Figure 3.4). He was an accomplished mathematician, later credited with some of the theorems of elementary algebra still taught today.

He was also active in promoting the atomic theory, which hypothesized that matter was composed of atoms. In a letter in 1607, he tried to persuade Kepler to adopt an atomic theory in his study of optics. Kepler declined, but a few years later in his *De Nive Sexangula* (1611) (Figure 3.5) he adopted an atomistic approach to describe the origin of the hexagonal shape of snowflakes. To do this he assumed that the snowflakes are composed of tiny spheres.[2]

His construction of the "most compact solid" is described as follows:

> For in general equal pellets, when collected in any vessel, come to a mutual arrangement in two modes according to the two modes of arranging them in a plane.

[1] Rukeyser, M., 1972, *The Traces of Thomas Harriott* (London: Gollanz).
[2] Leppmeier, M., 1997, *Kugelpackungen von Kepler bis heute* (Wiesbaden: Vieweg).

SEXANGVLA.                                    9

*Nam si errantes in eodem plano horizontali globulos aequales coegeris in angustum, ut se mutuo contingant, aut triangulari forma coeunt, aut quadrangulari; ibi sex unum circumstant, hic quatuor: utrinque eadem est ratio contactus per omnes globulos, demptis extremis. Quinquanguli forma nequit retineri aequalitas, sexangulum resolvitur in triangula: ut ita dicti duo ordines soli sint.*

*Iam si ad structuram solidorum quam potest fieri arctissimam progredieris. ordinesq; ordinibus superponas, in plano prius coaptatos,*

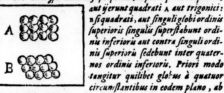

*aut yerunt quadrati A aut trigonici: B si quadrati, aut singuli globi ordinis superioris singulis superstabunt ordinis inferioris aut contra singuli ordinis superioris sedebunt inter quaternos ordinis inferioris. Priori modo tangitur quilibet globus a quatuor circumstantibus in eodem plano, ab uno supra se, & ab uno infra se: & sic in universum a sex aliis, eritq; ordo cubicus, & compressione facta fient cub: sed non erit arctissima coaptatio. Posteriori modo praterquam quod quilibet globus a quatuor circumstantibus in eodem plano tangitur. etiam a quatuor infra se, & a quatuor supra se, & sic in universum a duodecim tangetur; fientq; compressione ex globosis Rhombica. Ordo hic magis assimilabitur octaedro & Pyramidi. Coaptatio fiet arctissima: ut nullo praeterea ordine plures globuli in idem vas compingi queant. Rursum si ordines in plano structi fuerint trigonici; tunc in coaptatione solida aut singuli globi ordinis superi: superstant singuli inferioris, coaptatione rursum laxa, aut singuli superioris, sedent inter ternos inferioris Priori modo tangitur quilibet globus a sex circumstantibus in eodem plano, ab uno supra, & ab uno infra se, & sic in universum ab octo aliis. Ordo assimilabitur Prismati, & compressione facta fient pro globulis columna senum laterum quadrangulorum, duarumq; basium sexangularum. Posteriori modo fiet idem, quod prius posteriori modo in*

B            qua-

**Figure 3.5** The original page of *De Nive Sexangula* where the two regular packings of spheres in the plane are drawn (square [A] and triangular [B]).

If equal pellets are loose in the same horizontal plane and you drive them together so tightly that they touch each other, they come together either in a three-cornered or in a four-cornered pattern. In the former case six surround one; in the latter four. Throughout there is the same pattern of contact between all the pellets except the outermost. With a five-sided pattern uniformity cannot be maintained. A six-sided pattern breaks up into three-sided. Thus there are only the two patterns as described.

Now if you proceed to pack the solid bodies as tightly as possible, and set the files that are first arranged on the level on top of others, layer on layer, the pellets will be either in squares (A in diagram), or in triangles (B in diagram). If in squares, either each single pellet of the upper range will rest on a single pellet of the lower, or, on the other hand, each single pellet of the upper range will settle between every four of the lower. In the former mode any pellet is touched by four neighbors in the same plane, and by one above and one below, and so on throughout, each touched by six others. The arrangement will be cubic, and the pellets, when subjected to pressure, will become cubes. But this will not be the tightest packing. In the second mode not only is every pellet touched by it four neighbours in the same plane, but also by four in the plane above and by four below, and so throughout one will be touched by twelve, and under pressure spherical pellets will become rhomboid. This arrangement will be more comparable to the octahedron and pyramid. *This arrangement will be the tightest possible, so that in no other arrangement could more pellets be packed into the same container.*[3]

The structure here described by Kepler is cubic close packing, also called face-centered cubic (fcc). It has the greengrocer's packing fraction $\rho = 0.7404\ldots$. It is a crystalline structure: the local configurations repeat periodically in space like wall paper but in three dimensions. The regular repeating unit is a single sphere in this case. Each is surrounded by the same configuration with 12 neighbors in contact.

Kepler's work was the first attempt to associate the external geometrical shape of crystals with their internal composition of regularly packed microscopic elements. It was very unusual for his time, when the word "crystal" was applied only to quartz, which was thought to be permanently frozen ice.

Kepler asserted that the cubic close packing "will be the tightest possible, so that in no other arrangements could more pellets be packed into the same container." Despite Kepler's confidence this conjecture long resisted proof and became the oldest unsolved problem in discrete geometry.[4]

## 3.4 Marvelous Clarity, Neurotic Anxiety: The Life of Kepler

Kepler was born in Weil der Stadt (near Leonenberg, Germany) in 1571 (Figure 3.6). He was originally destined for the priesthood, but instead took up a position as school teacher of mathematics and astronomy in Graz.

---

[3] Quoted from: Kepler, 1966, *The Six-Cornered Snowflake*, translated by C. Hardie (Oxford: Clarendon).

[4] Discrete geometry concerns geometrical objects and properties where the study does not essentially rely on the notion of continuity.

**Figure 3.6**   Johannes Kepler (1571–1630).

When Kepler arrived in Graz he was 25-years-old and much occupied with astrology. He issued a calendar and prognostication for 1595 which contained predictions of bitter cold, a peasant uprising and invasions by the Turks, which was probably par for those times. All were fulfilled, greatly enhancing his local reputation.

Kepler was an enthusiastic Copernican. Today, he is chiefly remembered for his three laws on planetary motion but his search for cosmic harmonies was much broader, ranging from celestial physics to sphere packings.

Kepler's personality has been described as "neurotically anxious." Certainly he had an unhappy personal life. The story goes that, in seeking to optimize the partner for his second marriage, he carefully analysed the merits of no less than 11 girls—before choosing the wrong one. The word "nerd" may be inappropriate for a giant of the Renaissance but it springs to mind.

On the other hand Kepler's scientific writing presents us with what he called a "holy rapture" of compelling power. One young scientist in the twentieth century who was inspired by its fusion of scientific insight and religious mysticism was L. L. Whyte.[5] This is his translation of the breathless cadenza from Kepler's *Harmony of the Worlds*.

But now, since eighteen months ago the first light dawned, since three moons the full day, and since a few days the sunshine of the most marvellous clarity—now nothing holds me back: now

---

[5] Whyte, L. L., 1963, *Focus and Diversions* (London: Cresset Press).

I may give in to this holy rapture. Let the children of men scorn my daring confession: Yes! I have stolen the golden vessels of the Egyptians to build from them a temple for my God, far from the borders of Egypt. If you forgive me, I am glad; if you are angry I must bear it. So here I throw my dice and write a book, for today or for posterity. I care not. Should it wait a hundred years for a reader, well, God himself has waited six thousand years for a man to read his work.

Whyte suggested that there should be a verb "to kepler, meaning to identify a conceivable form of order as an aim of search." Perhaps it could have the secondary meaning "to overidealize real systems, in an attempt to scientifically analyse them," as in Kepler's search for a wife. Recent attempts of physicists to understand the housing market may fall into this category.

## 3.5   Progress by Leaps and Bounds?

While the Kepler problem remained unsolved, many mathematicians contributed to this study by offering something less than the full theorem that is required.

Gauss demonstrated that the face-centered cubic structure is the densest *lattice*[6] packing in three dimensions. But this is not sufficient since denser local configurations exist (such as the tetrahedral configuration of four spheres mutually in contact) and therefore noncrystalline structures could conceivably have packing fractions higher than that of the cubic close packing.

Fejes Tóth reduced the problem to a finite but impossibly large calculation: "It seems that the problem can be reduced to the determination of the minimum of a function of a finite number of variables."[7]

Often mathematicians set themselves the task of proving that the highest density must be lower than some value $X$. This is an *upper bound*. With no thought at all we can offer $X = 1.0$ for an upper bound of $\rho$, and with some subtlety much better values may be found. Obviously, if we could show that 0.7404... is an upper bound, then, we would also know we can reach it by the method of the greengrocer. The ball game would be over, apart from questions of uniqueness.

A particularly nice and simple bound is that of C. A. Rogers (1958); it is precisely the packing fraction that we recognized as appropriate to a

---

[6] The special case of a crystalline structure in which every sphere has an equivalent position.

[7] Fejes, Tóth L., 1964, *Regular Figures* (New York: Pergamon).

perfect tetrahedral packing (which cannot exist), that is, 0.7797.... Better bounds have followed: 0.778 44 ... (Lindsey 1987), 0.778 36 ... (Muder 1988), 0.7731 (Muder 1993) and others. There are older ones as well: 0.828 (Rankin 1947), 0.883 ... (Bichfeldt 1929).[8]

This was progress by bounds but hardly by leaps. A proof of the original proposition still seemed far off, until recent times.

---

[8] Conway, J. H. and Sloane, N. J. A., 1988, *Sphere Packings, Lattices and Groups* (Berlin: Springer).

# Chapter 4

## Proof Positive?

### 4.1   News from the Western Front

It may be safely assumed that quite a few experts have devoted some small fraction of their time to looking for a solution to the Kepler problem, rather as the punter places a small bet on a long-odds horse. One would not want to stake a whole career on it, but the potential rewards are attractive enough to compel attention.

The problem was included in a celebrated list drawn up by David Hilbert at the dawn on the twentieth century, which was like a map of the mathematical universe for academic explorers and treasure hunters. We have already cited his challenge to posterity in Chapter 1. Some of Hilbert's treasures have been found from time to time, but the key to Kepler's conjecture lay deeply buried. As D. J. Muder said, "It's one of those problems that tells us that we are not as smart as we think we are."

In 1991 it seemed that the key had finally been found by Wu-Yi Hsiang. The announcement of the long awaited proof came from the lofty academic heights of Berkeley, California. Hsiang had been a professor there since 1968, having graduated from Taiwan University and taken a Ph.D. at Princeton. Few American universities enjoy comparable prestige, so the mathematical community was at first inclined to accept the news uncritically. When Ian Stewart told the story in 1992 in his *Problems of Mathematics*, he described Hsiang's work in heroic terms, but wisely added some cautionary touches to the tale.

Many mathematical proofs are long and involved, taxing the patience of even the initiated. There has to be a strong element of trust in the early acceptance of a new theorem. So the meaning of the word *proof* is a delicate philosophical and practical problem. The latest computer-assisted proofs have redoubled this difficulty. In the present case, most of the mathematics

---

### A Comment by Kantor on Hilbert's 18th Problem

Hilbert's text gives the impression that he did not anticipate the success and the developments this problem would have.

The hexagonal packing in plane is the densest (proof by Thue in 1892, completed by Fejes in 1940). In space, the problem is still not solved. Although there is very recent progress by Hales. For spheres whose centers lie on a lattice, the problem is solved in up to eight dimensions. The subject has various ramifications: applications to the geometry of numbers, deep relations between coding theory and sphere-packing theory, the very rich geometry of the densest known lattices. (Kantor, J. M., 1996, *Math. Intelligencer*, Winter, p. 27.)

---

was of an old-fashioned variety, close to that of school-level geometry. Indeed, Hsiang claimed that a retreat from sophistication to more elementary methods was one secret of his success. Nevertheless, his preprint ran to about 100 pages and was not easily digested even by those hungry for information. As his colleagues and competitors picked over the details, some errors became apparent.

This is not unusual. Another recent claim, of an even more dramatic result—the proof by Wiles[1] of Fermat's Last Theorem—required some running repairs, but is still considered roadworthy and indeed prizeworthy. However, Hsiang did not immediately succeed in rehabilitating his paper.

Exchanges with his critics failed to reach a resolution. A broadside was eventually launched at Hsiang by Thomas Hales (Figure 4.1) in the pages of the splendidly entertaining *Mathematical Intelligencer*. Hales' piece lies at the serious end of that excellent magazine's spectrum but nevertheless it grips the reader with its layers of implication and irony, most unusual in a debate on a piece of inscrutable academic reasoning.

Despite the inclusion of some conciliatory gestures ("promising programme," "improves the method") the overall effect is that of a Gatling gun, apparently puncturing the supposed proof in many places. Hales begins with the statement that "many of the experts in the field have come to the conclusion that [Hsiang's] work does not merit serious consideration" and ends with a demand that the claim should be withdrawn: "Mathematicians can easily spot the difference between hand-waving and proof."

---

[1] Wiles, A., 1995, Modular elliptic-curves and Fermat's Last Theorem, *Ann. Math.* **141** 443–551.

**Figure 4.1** Thomas Hales.

Hsiang replied at length in the same magazine, protesting against the use of a "fake counter example." Meanwhile, Hales and others were themselves engaged in defining programmes for further work, as the explorer stocks supplies and makes sketch-maps for a hopeful expedition. Indeed, he was already at base camp.

## 4.2 The Programme of Thomas Hales

Mathematicians often speak of what European English would call a "programme," upon whose execution they are engaged, and do not mean a computer *program*. Rather, it is a tactical plan designed to achieve some objective. Like mountaineers who wish to conquer Everest, they define in advance the route and various camps which must be established along the way. It is in this spirit that Thomas Hales has attacked the Kepler problem. He had established base camp when we set out to write this book. He had proved a number of intermediate results and felt that the summit could be reached.

Hales' programme was based on reducing the Kepler problem to certain *local* statements about packing. Earlier we pointed to the impossibility of packing regular tetrahedra (Chapter 3), and saw the Kepler problem as one of "frustration," for this reason. But this does not mean that the global packing problem cannot be reduced to a local one, in some more subtle sense.

Hales considers a *saturated* packing, which is an assembly of noninter-secting spheres within which no further nonintersecting spheres can be added. He uses *shells*: local configurations made of a sphere and its surrounding neighbors. He calculates the local density and a *score*, this quantity being associated with the empty and occupied volume in the local packing around the sphere. The "programme" of Hales was to prove that all the possible local configurations have scores lower or equal than 8.0, that is, the one associated with Kepler's dense packing. This is enough to prove Kepler's conjecture.

Hales started the implementation of this programme around 1992. He soon proved that a large class of local packings score less than 8 but there remained a few local configurations for which this proof was extremely tricky. In spring 1998, one could read on his home page on the Internet: "When asked how long all this will take, I leave myself a year or two. But I hope these pages convince you that the end is in sight!" He estimated that it would take until year 2000 to finish the proof.

The main problem in this kind of proof is to find a good way to partition a packing into local configurations. This is a key issue: How does one properly define "shells" in the packing? All the major breakthroughs in the history of the Kepler conjecture (including the Hsiang attempt) are associated with different ways of partitioning space. In particular, there are two natural ways to divide the space around a given sphere in a packing.

The first is the *Voronoï decomposition* (to be further described in Chapter 7). The Voronoï cell is a polyhedron, the interior of which consists of all points of the space which are closer to the center of the given sphere than to any other. This was the kind of decomposition adopted by Fejes Tóth to reduce the Kepler problem to the "determination of the minimum of a function of a finite number of variables" and to settle the 2D coin-packing problem (Appendix A). But this method meets with difficulties for some local configurations, such as when a sphere is surrounded by 12 spheres with centers on the vertices of a regular icosahedron. In this case the local packing fraction associated with the Voronoï cell is $\rho = 0.7547\ldots$ which is bigger than the value of the Kepler packing ($\rho = 0.7404\ldots$) and its score is bigger than 8. We are frustrated.

The second natural way of dividing space is the *Delaunay decomposition*. Here space is divided in Delaunay simplexes, which are tetrahedra with vertices on the centers of the neighboring spheres chosen in a way that no other spheres in the packing have centers within the circumsphere of a Delaunay simplex. The local configuration considered is now the union of the Delaunay tetrahedra with a common vertex in the center of a given sphere (this is called the *Delaunay star*). This was the kind of decomposition first adopted by Hales in his programme. For instance, the Delaunay decomposition succeeds in the icosahedral case, giving a score of 7.999 98. But there exists at least one local configuration with a higher score.

**Figure 4.2** The assembly of 12 spheres around a central one in a pentahedral prism configuration.

This is an assembly of 13 spheres around a central one, which is obtained by taking 12 spheres centered at the vertices of an icosahedron and distorting the arrangement by pressing the thirteenth sphere into one of the faces. This configuration scores 8.34 and has local packing fraction $\rho = 0.7414$. Another nasty configuration which has a score dangerously close to 8 is the "pentahedral prism." This is an assembly of 12 spheres around a central one, shown in Figure 4.2.

The Voronoï and Delaunay decompositions can be mixed in infinitely many ways. This is what Hales attempted by decomposing space in "Q-systems" and associated stars. This decomposition was successful to establishing the score of the pentahedral prism at 7.9997. However, new and nasty configurations remained to be considered.

## 4.3 At Last?

On August 10, 1998, as one of the authors of this book was picking his fishing tackle out of the back of his car, his eye fell on a headline in a British newspaper. All thoughts of angling were dismissed for a while.

KEPLER'S ORANGE STACKING PROBLEM QUASHED

In a short report Simon Singh announced Thomas Hales' success, and quoted John Conway, a leading expert and mentor of Hales: "For the last decade, Hales' work on sphere packings has been painstaking and credible. If he says he's done it, then he's quite probably right."

Back at the office an e-mail message had been received, which must have disturbed quite a few other summer holidays as well.

```
From        Thomas Hales
Date:       Sun,    9 Aug   1998      09:54:56
To:
Subject: Kepler conjecture
```

Dear colleagues,

I have started to distribute copies of a series of papers giving a solution to the Kepler conjecture, the oldest problem in discrete geometry. These results are still preliminary in the sense that they have not been refereed and have not even been submitted for publication, but the proofs are to the best of my knowledge correct and complete.

Nearly four hundred years ago, Kepler asserted that no packing of congruent spheres can have a density greater than the density of the face-centered cubic packing. This assertion has come to be known as the Kepler conjecture. In 1900, Hilbert included the Kepler conjecture in his famous list of mathematical problems.

In a paper published last year in the journal `Discrete and Computational Geometry', (DCG), I published a detailed plan describing how the Kepler conjecture might be proved. This approach differs significantly from earlier approaches to this problem by making extensive use of computers. (L. Fejes Toth was the first to suggest the use of computers.) The proof relies extensively on methods from the theory of global optimization, linear programming, and interval arithmetic.

The full proof appears in a series of papers totalling well over 250 pages. The computer files containing the computer code and data files for combinatorics, interval arithmetic, and linear programs require over 3 gigabytes of space for storage.

Samuel P. Ferguson, who finished his Ph.D. last year
at the University of Michigan under my direction,
has contributed significantly to this project.

The papers containing the proof are:

An Overview of the Kepler Conjecture, Thomas C. Hales
A Formulation of the Kepler Conjecture, Samuel P.
Ferguson and Thomas C. Hales

```
Sphere Packings I,      Thomas C. Hales
                        (published in DCG, 1997)
Sphere Packings II,     Thomas C. Hales
                        (published in DCG, 1997)
Sphere Packings III,    Thomas C. Hales
Sphere Packings IV,     Thomas C. Hales
Sphere Packings V,      Samuel P. Ferguson
The Kepler Conjecture (Sphere Packings VI),
                        Thomas C. Hales
```

Postscript versions of the papers and more information
about this project can be found at

http://www.math.lsa.umich.edu/~hales

Tom Hales

A month later, after Hales had enjoyed his own holiday, he kindly con-
sented to answer a few e-mailed questions that might illuminate this account
of his achievement. His answers were as follows, with some minor editing:

```
Date: Wed, 30 Sep 1998    08:42:12
From: Tom Hales
Reply-To: Tom Hales
To:
cc:
Subject: Re: your mail

>       Dear Thomas
>
>       - when were you first attracted to the problem?
```

In 1982, I took a course from John Conway on groups
and geometry.

>      - what was the hardest part?

The problem starts out as an optimization problem in
an infinite number of variables. The original problem
must be replaced by an optimization in a finite number
of variables. It was extremely difficult to find a
finite-dimensional formulation that was simple enough
for computers to handle.

>      - in what your method is different from the
>           previous approaches?

This approach makes extensive use of computers,
especially interval arithmetic and linear programming
methods. Most previous work was based on the Voronoi
cells. This approach creates a hybrid of Voronoi
cells and Delaunay simplices.

>      - did you follow the lead/style of anyone in
>           particular?

My greatest source of inspiration on this problem was
L. Fejes Toth. He was the first to propose an
optimization problem in a finite number of variables
and the first to propose the use of computers. But my
proof differs from the program he originally proposed.

>      - is it correct to say that this is a
>           "traditional" proof with no
>           significant elements of computer-based
>           proof?

Not at all. The computer calculations are an essential
part of the proof.

>      - what was your first reaction when Hsiang
>           claimed a proof?

I have followed his work closely from the very start.
My doubts about his work go back to a long discussion
we had in Princeton in 1990. I discuss my reaction to
his work further in my Intelligencer article.

>       - have you always been confident in success?

In the fall of 1994, I found how to make the hybrid decompositions work, combining the Voronoi and Delaunay approaches. I have been optimistic since then.

>       - your proof fill 10 papers and about 250
>         pages, why does it need so much?

This is a constrained nonlinear optimization problem involving up to 150 variables with many local maxima that come uncomfortably close to the global maximum. Rigorous approaches to problems of this complexity as generally regarded as hopelessly difficult.

>       - do you think that in the future a different
>         approach might be able to reduce the size
>         and complexity of this proof?

This is not an optimal proof. I have concrete ideas about how the proof might be simplified. Although I'm quite certain the proof can be simplified, it will require substantial research to carried this out. There could also be other proofs along completely different lines, but I do not have any definite ideas here.

>       - can you calculate the maximum size of a
>         cluster of spheres with a given density
>         (larger than the Kepler one)?

This is an interesting question that a number of people would like to understand. I would be curious to know whether my methods might lead to something here.

## 4.4   Polishing Off the Programme

It immediately appeared likely that Hales would be accorded the credit for the definitive solution of the Kepler Problem. In January 1999, a special conference was devoted to understanding the proof. A panel of 12 referees, headed by Gabor Fejes Tóth,[2] set to work on checking the proof. Six years later, Hales was ready to assemble the complete proof in *Annals of Mathematics*, in an "abridged" version of well over 100 pages.[3] "Some uncertainty remained," in the words of the panel, but publication proceeded. A report from the editor admitted:

> The news from the referees is bad, from my perspective. They have not been able to certify the correctness of the proof, and will not be able to certify it in the future, because they have run out of energy to devote to the problem. This is not what I had hoped for.

This was followed in 2006 by six papers, leaving no stone unturned. Many of the historical points that we have touched upon are briefly and authoritatively reviewed in the first of these papers.[4] Hales begins the story of cubic close packing rather earlier than we have done, with a reference to a work in Sanskrit, dated 499 C.E.

His final words in the sixth paper are generally accepted today: "This completes the proof of the Kepler conjecture." Thus ends a chapter in mathematics that one of its practitioners has called "scandalous." This refers not to the Hsiang's previous venture but rather to the general embarrassment of mathematicians over several centuries.

## 4.5   The Acceptance of Proof

We have seen that the acceptance of a deep proof is not a straightforward matter. To form a judgment requires arcane knowledge and great commitment, and much time. Kepler himself said he would gladly wait a hundred years for a reader. But if confronted by one of today's committees to determine promotion to academic tenure, he might not have been so sanguine.

Mathematics confronts a difficulty here, but then so do those branches of experimental science in which procedures are so complex as to be hard to reproduce, as Fejes Tóth remarked in his report for *Annals of Mathematics*. A substantial amount of new work has to be taken on trust, even though trust is an increasingly rare commodity.

---

[2] The son of László Fejes Tóth is mentioned several times in this book.

[3] Hales, T. C., 2005, A proof of the Kepler conjecture, *Annals of Mathematics* **162**, 1065–1085.

[4] Hales, T., 2006, *Discrete and Computational Geometry* **36**, 5.

The public has been awakened to the dramatic potential of this process, when compounded with the frailties and oddities of human nature. David Auburn's play "Proof" has been filling theaters on Broadway and elsewhere since 2000. Appropriately, Auburn combines a knowledge of mathematics with a grounding in political philosophy.

At the time of writing, a real-life drama—a resolute one-man show—is being acted out by the Grigori Perelman, who has announced (but not fully divulged) a proof of the Poincaré Conjecture. So it is that mathematical proof has reached the columns of the *New Yorker*, and a thousand blogs besides.

A further dimension has been added to the fascination of proof by the use of the computer, as in the case of the work of Hales.

Already in 1967, the announcement in the *New York Times* of a computer proof of the four-color conjecture by Kenneth Appel and Wolfgang Haken sparked off a lively debate on the acceptability of such methods. Forty years later, they are much less controversial, as we continue to accommodate ourselves to the ever greater role of computers in our lives. It would be more provocative to debate, say, the limits of artificial intelligence or the possibilities of computer consciousness than to dispute the validity of computer proofs. Nevertheless, since we have encountered them in packing theorems, particularly in the work of Hales, let us rehearse the arguments as presented by, for example, Davis and Hersh in *The Mathematical Experience* (1981).

The traditional view has held that it must be possible to check every step in a proof, in order for it to be acceptable. In some cases, such as that of the proof of Fermat's Last Theorem by Wiles, few will ever be able to accomplish this. Those that do so (who may include the referees for publication, if they are conscientious) will be the guarantors upon which the rest of us can rely. But modern computer proofs generate such a multitude of logical operations within the machine that it is beyond any human capacity to follow them, even if they are made manifest.

There is no getting away from the necessity to rely on the correct functioning of the machine and its software, in response to the programmer's instructions. They can be checked only at the level of programming, or in terms of the reproducibility of the result on a quite different machine, rather as some experimental results are tested. If these precautions are taken, today's generation is quite happy to accord the same status to the proof as in days of old. They will say: There was always a tiny element of uncertainty in any elaborate proof of the more traditional kind. Human beings are susceptible to error, as are computers. Rather more so, perhaps?

Some of us would still sigh. We point to economy, elegance and transparency as cardinal virtues of a *good* mathematical proof, award low scores to the new methodology on those criteria, and call for renewed efforts to be more explicit.[5]

Perhaps some future flash of insight will replace the labors of Hales by something more direct and economical. And mathematicians always aspire to produce a proof so elegant as to be worthy of inclusion in the mythical book of God, whimsically conceived by Paul Erdös, in which only the most exquisite are to be found.

## 4.6 The Flyspeck Project

In the meantime, the approach of Hales would appear to be "the only game in town." Rather than pine for the old days, might we just try to do a more fully convincing job of Hales' proof, given its importance? This is the motivation of the Flyspeck Project, in which Hales himself is involved. According to the project's Web page:

> The purpose of the flyspeck project is to produce a formal proof of the Kepler Conjecture. The name "flyspeck" comes from matching the pattern /f.*p.*k/ against an English dictionary. FPK in turn is an acronym for "The Formal Proof of Kepler." The term "flyspeck" can mean to examine closely or in minute detail; or to scrutinize. The term is thus quite appropriate for a project intended to scrutinize the minute details of a mathematical proof.[6]

The very nature of the project is an indication of changing times. What is called for here is not mere "formality" in the colloquial sense. It demands a logical structure in which *nothing* is assumed or omitted.

The layman might ask: is that not what mathematics is supposed to be? In principle it is, but complete rigor (in a logical sense) has been left to philosophers, while mathematicians have agreed on what may be reasonably left unsaid.

The Flyspeck Project is of even more heroic proportions than the original effort of Hales. Twenty work-years (whatever a "work-year" is) are foreseen. *Qui vivra verra.* . . .

Flyspeck is resolutely based on computer programming: the ultimate goal is to "design a computer program that produces carefully checked mathematical theorems." Will God approve?

---

[5] For a further exploration of this ongoing debate, see the article by: Horgan, J., 1993, The death of proof, *Sci. Am.* October **75**, and the responses and discussion that followed in later issues.

[6] http://www.math.pitt.edu/thales/flyspeck/.

## 4.7   Who Cares?

*Who cares* is a fair question, often addressed to startled scientists and mathematicians by puzzled journalists. Particularly in this case, why does proof matter, whenever we are sure we know the truth, anyway?

One answer is that some people did admit to a tiny sliver of doubt insinuating itself into the certainty of their conviction. These were usually not physical scientists, who would assert that if there was a better structure they would have spotted it by now, written somewhere in the book of nature. This may be insufficiently humble: surprises do occur from time to time, even in crystallography. They often result in Nobel Prizes. As Alfred North Whitehead said, "In creative thought common sense is a bad master. Its sole criterion for judgment is that the new ideas shall look like the old ones, in other words it can only act by suppressing originality."

Mathematics does not consist entirely of theorems. There is a rough texture of conjecture and useful approximation, held up here and there by a rigorous proof, serving the same purpose as the concrete framework of a building. The more of these the better, to stop the whole thing collapsing under the weight of loose speculation.

Rigorous proofs and exact results are like the gold bars in the vaults of the Federal Reserve, guaranteeing the otherwise unreliable monetary transactions of the world. They are apparently useless since they are not put to any direct use, and yet they have real value. (Since the first draft of this book was written there has been much debate on the abandonment of gold reserves, so the simile may soon be a feeble one. The value of a proof is more durable.)

## 4.8   The Power of Thought

In one of many important articles on polyhedra and packings, the eminent Canadian mathematician Coxeter chose to begin with a poem by Charles Mackay:

> *Cannonballs may aid the truth,*
> *But thought's a weapon stronger,*
> *We'll win our battles by its aid;*
> *Wait a little longer.*

To this we might add a verse to bring it up to date:

> *If you fail to reach your end,*
> *your siege has come to nought,*
> *call on your electronic friend,*
> *to spare a microsecond's thought.*

Whether computers can really "think" is a deeper problem than the merely technical one of proof-checking. We may someday accept an even stranger role for these machines in mathematics. Will they write books as well? They already print and sell them, so this may be a case for vertical integration.

# Chapter 5

# Disordered Packings

## 5.1  Balls in Bags

Even perfect spheres do not readily fall into the closest possible packing when, for example, gently shaken in a bag. This may be regarded as a practical consequence of the characteristic of frustration which is at the heart of the difficulty of the Kepler Problem. The local density can exceed the global optimum of cubic close packing. Shaken in a bag, how are spheres to "know" this, and find the right global compromise?

Many other kinds of hard particles (sand, powder, pebbles) of less regular size and shape, are even less likely to find an ordered state. So the disordered packing of spheres can stand as the prototype for a wide class of important granular materials. It is worth pursuing with some precision. We do not have to go all the way with Cyril Stanley Smith, for whom ordered states were "dead" and disordered ones "alive," to appreciate their importance.

Let us therefore look more closely at the ball bearings in the bag. In order to fix their positions, wax may be poured in and the contents then dissected. The first person to undertake this experiment systematically was J. D. Bernal, or rather his students Finney and Mason, in the 1950s.[1] The random packing of balls became known as Bernal packing (Figure 5.1). One might wonder why it was not done before. Certainly it is tedious, but such tedium is often part of the price of a Ph.D.

The preceding century, in which the detailed atomic arrangements of crystals were hypothesized and powerful theories of symmetry applied to them, was one in which perfect order was the ideal (as one might expect

---

[1] Bernal, J. D., 1959, A geometrical approach to the structure of liquids, *Nature* **183**, 141–147.

(a)                                    (b)

**Figure 5.1**  Bernal packing of spheres.

in an imperial age). Order and beauty were often interchangeable in the sensibility of the admirer of nature.

*Là, tout n'est qu'ordre et beauté*

Charles Baudelaire (1821–1867)

At the century end, as the old order fell into decay, the mood began to change. The poet Hopkins gloried in "dappled things," the commonly observed disarray of the real world. Nevertheless, words like "impurity" and "defect," applied to departures from perfect order, still carried a prejudicial overtone in physics, even as they eventually emerged as the basis of the semiconductor industry.

Those whose curiosity centered on the structure of liquids still tended to picture them as defective crystals or construct elegant formal theories that had little to do with their characteristically random geometry.

Eventually Bernal, perhaps because of his biological interests, saw the necessity to examine this geometry more explicitly, to confront and even admire its variety. He also recognized the difficulty of doing so, other than by direct observation of a model system. Why not ball bearings? He called this unsophisticated approach "a new way of looking" at liquids.

Hence, the shaking and kneading of ball bearings in a bag, the pouring in of wax, the meticulous measurement of the positions of balls as the random cluster was disassembled. The packing fraction of the Bernal packing was found to be roughly 0.64; or slightly less if the balls are not kneaded to

**Figure 5.2**    Desmond Bernal (1901–1971).

encourage them to come together closely. Although particular local arrangements recur within it, they are variable in shape and random in distribution.[2]

Subsequently, the Bernal structure has found applications in models of the structure of amorphous metals and colloids that consist of hard spherical particles.

> *Si l'ordre satisfait la raison,*
> *le désordre fait les délices de l'imagination.*

Paul Claudel (1868–1955)

## 5.2    A New Way of Looking

Desmond Bernal (Figure 5.2), born in 1901 on an Irish farm, was one of the prime movers of modern crystallography and biophysics. He is thought to have narrowly missed a Nobel Prize for his work on sex hormones. This would have been singularly appropriate for a man whose sexual appetite was rumoured to be prodigious. This is unusual in a scientist, as Bernal himself found out when curiosity moved him to check the historical record for the exploits of others. He found that typically they had less adventurous

---

[2] Bernal, J. D. and Mason, J., 1960, Co-ordination of randomly packed spheres, *Nature* **385**, 910–911.

erotic *curricula vitae* than his. Certainly, few can have enjoyed the excitement of being pursued down the street by a naval officer with a revolver after a brusque interruption of an amorous interlude.

His intellectual brilliance is better documented. It would have found more positive expression if it had not been deflected into political channels (he was an ardent Communist—some of his sociopolitical writings are still highly regarded). Despite such distractions, he gathered around him in London an outstanding international research group.

Among the thoughts that most fascinated him from the outset were ideas of packing. They eventually found expression in his study of the structure of liquids, which proceeded directly along the down-to-earth lines advocated by Lord Kelvin, the founder of the "hands-on" school of British crystallography.

Bernal used his hands and those of his students to build large models or take apart packings of ball bearings. He showed that geometrical constraints impose organization and local order on such random structures.

## 5.3 How Many Balls in the Bag?

The Bernal packing falls well short of the maximum density that can be achieved in an ordered packing. Nevertheless, it has acquired its own significance as the best *random* packing. It is difficult to give to this notion any precise meaning, but many experiments and computer simulations of different kinds do reproduce the same value (to within a percent or so). For example, spheres may be added to a growing cluster, according to various rules, and the eventual result is a random packing not very different from that of Bernal. This was the finding of Charles Bennett, who wrote various computer programs for such "serial deposition" at Harvard University in the mid-1960s. (Bennett has gone on to be one of the most authoritative and imaginative theoreticians in the science of computation.)

From the outset, this was called *random close packing*, to distinguish it from looser packings of spheres which were found in some experiments. Attempts to define a unique density for random *loose* packings are probably futile, because it must depend on the precise circumstances, and physical effects (such as friction) which contribute to it. For instance, G. D. Scott poured thousands of ball bearings into spherical flasks of various sizes.[3] When the flask was gently shaken to optimize the packing, the density was found to be $\rho = 0.6366 - 0.33N^{-1/3}$, with $N$ being the number of balls. When the flask was not shaken, the loosest random packing was found to have $\rho = 0.60 - 0.37N^{-1/3}$. Lower values for the packing fraction can be

---

[3] Scott, G. D., 1960, Packing of spheres, *Nature* **188**, 908–909; 1969, *Brit. J. Appl. Phys.* **2**, 863.

obtained by eliminating the effect of gravity. The lowest densities which can be experimentally obtained are around $\rho \simeq 0.555$.[4]

## 5.4   Is the Bernal Close Packing Well Defined?

Since Bernal's work there has been a very large number of simulations and experiments, all consistently concluding that when equal spheres are packed in a disordered fashion the packing fraction does not exceed ∼0.64. Therefore the "Bernal close packing limit" is—empirically—a very well defined bound. However, from a fundamental perspective it is not clear what actually happens at such packing fraction. The problem is intrinsically related to disorder. The Kepler packing consists in only one configuration repeated in space, and here instead we have a very large variety of local configurations which combine in a even larger variety of packings. The question is: What do all these packings have in common?

The answer might come from the equivalently complex domain of the studies of glasses. The idea is that at that limit the system exhaust its capability of changing configurations and becomes frozen in a state of permanently zero "granular temperature" (see Section 6.7). Scott's work tells us that the final density depends slightly on how we treat the sample. Gentle shaking is necessary to allow it to settle down with maximum density. For some this shaking treatment is analogous to the finite temperature fluctuations to which atoms are subject. Shaking (or tapping) with gradually decreasing force is likened to lowering the temperature.

## 5.5   Bernal's Long-Running Ball Game

Half a century ago, Bernal took a close look at random sphere packings, as we have recounted. He recognized recurring local arrangements of which the most common were the tetrahedron and the bipyramid.

In the random packing these are distorted from their ideal symmetric shapes. Some have wondered whether Bernal, in trying to set up a taxonomy of local geometry based on these, was clutching at straws, or "keplering." Nevertheless, his point of view has been adopted by many others and most recently by a Russian team investigating the signature of structural changes at the Bernal dense packing limit.[5] They concentrated on the tetrahedra that are such a pervasive motif in this book, and excluded from consideration those that are heavily distorted (according to an arbitrary rule). The number of tetrahedra increased as the Bernal dense

---

[4] Onoda, G. Y. and Linger, E. G., 1990, Random loose packing of uniform spheres and the dilatancy onset, *Phys. Rev. Lett.* **64**, 2727–2730.

[5] Anikeenko, A. V. and Medvedev, N. N., Polytetrahedral nature of the dense disordered packings of hard spheres, *Phys. Rev. Lett.* **98**, 235504.

packing was approached, and the tetrahedra were locked together in large clusters that included almost all of the spheres.

This vision of the significance of the Bernal dense packing is still a tentative one. Perhaps the random packing will forever escape the force of logic. But mathematical physicists do not give up easily.

## 5.6　Tomography Takes Over

We have recalled the labor-intensive examination of packings by dismantling them, in a manner akim to anatomical dissection. To get a precise picture of the packing, Scott poured molten paraffin wax in a heated container

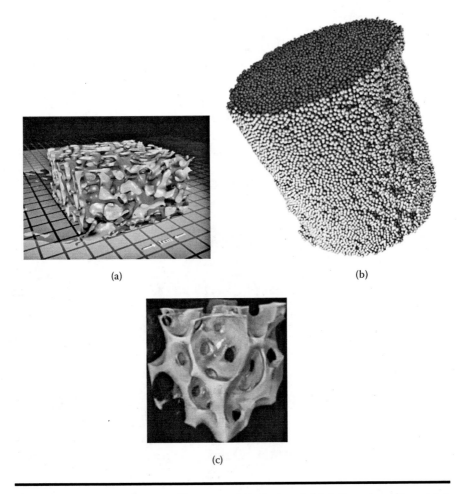

(a)　　　　　　　　　　　　　　　　　　　(b)

(c)

**Figure 5.3**　(a) A termite nest. (Courtesy of R. Corkery.) (b) A Bernal packing. (c) A detail of the inter granular space, all reconstructed from XCT tomography. (Courtesy of ANU XCT Group.)

filled with balls and by carefully dissecting it he measured the coordinates of each ball with a remarkable precision of less than 1% of their diameters. Not surprisingly this was the last attempt to measure the sphere position in Bernal packings until very recently. Now, all Scott's efforts are no longer necessary.

Instead we may look inside packings, and other dense structures, using x-rays[6] (Figure 5.3). In science their first application was to crystallography, to determine the arrangement of atoms by diffraction (Chapter 13), but their application to the human body aroused more interest and alarm. The kind of picture produced in a medical x-ray can be used for our packing as well, but with an additional twist that owes its practicality to the computer.

X-ray tomography requires a large number of pictures taken from different angles, from which a full *three-dimensional* representation can be constructed. Other alternatives such as magnetic resonance imaging (MRI), are also available, so that we can now truly look inside things, in full detail.

[6] Aste, T., Saadatfar, M. and Senden, T. J., 2005, Geometrical structure of disordered sphere packings, *Phys. Rev. E* **71**, 061302.

# Chapter 6

# Sands and Grains

## 6.1 Granular Materials

We have told the story of packing of identical spheres. Now, we adopt a wider perspective—that of *granular materials*. This term refers to systems of particles large enough to be unaffected by thermal motion, which is the normal state of affairs in our macroscopic world. For this to be the case, the typical grain size should be 1 micrometer or more, but the largest such "grains" could be of tectonic scales. It is estimated that we use (and mostly waste) 40% of our energy in handling and processing these materials which surround us in our daily lives. They are central to a very wide range of industries, from agricultural to pharmaceutical. Despite such a crucial role in most fields of human activity and scientific interest dating back to the nineteenth century, their properties remain elusive and difficult to capture in theory.

Granular materials can flow like liquids under some conditions, yet remain rigid under others.[1] Their intrinsic complexity is a consequence of the fact that they are composed of many pieces that can assemble in large structures that move (or refuse to move) collectively in surprising ways. Understanding this requires new paradigms and tools that go beyond the traditional domains of solid state physics, engineering and material science. It is forcing us to rethink established classifications of matter.

---

[1] van Hecke, M., 2005, Granular matter: A tale of tails, *Nature* **435**, 1041–1042; 2005, *Nature* **436**, 37; Buchanan, M., 2003, Think outside the sandbox, *Nature* **425**, 556–557; Umbanhowar, P., 2003, Granular materials—Shaken sand—A granular fluid? *Nature* **424**, 886–887.

**Figure 6.1** Osborne Reynolds (1842–1911).

## 6.2 Osborne Reynolds: A Footprint in the Sand

> It is seldom left for the philosopher to discover anything which has a direct influence on pecuniary interests; and when corn was bought and sold by *measure*, it was in the interest of the vendor to make the interstices as large as possible, and of the vendee to make them as small (...).
>
> If we want to get elastic materials light we shake them up (...) to get these dense we squeeze them into the measure. With corn it is the reverse; (...) if we try to press it into the measure we make it light—to get it dense we must shake it—which, owing the surface of the measure being free, causes a rearrangement in which the grains take the closest order.[2]

With these words Osborne Reynolds (Figure 6.1) described this "paradoxical" property of granular packings.

When granular material, such as sand or rice, is poured into a jar its density is relatively low and it flows rather like an ordinary fluid. A stick can be inserted into it and removed again easily. If the vessel, with the stick inside, is gently shaken the level of the sand decreases and the packing

---

[2] Reynolds, O., 1886, Experiment showing dilatancy, a property of granular material, possibly connected with gravitation, *Proc. Royal Institution of Great Britain* **11**, 354–363; Read February 12.

density increases. Eventually the stick can no longer be easily removed and when raised it will support the whole jar. This conveys a strong sense of the ultimate jamming together of the grains, as described by Bernal.

Such procedures have been the subject of research in physics laboratories in recent years,[3] but they can be traced back to the words of Osborne Reynolds.

> At the present day the measure for corn has been replaced by the scales, but years ago corn was bought and sold by measure only, and measuring was then an art which is still preserved. (...) The measure is filled over full and the top struck with a round pin called the strake or strickle. The universal art is to put the strake end on into the measure before commencing to fill it. Then when heaped full, to pull the strake gently out and strike the top; if now the measure be shaken it will be seen that it is only nine-tenths full.

The subtlety of compaction of granular systems has been long known and Jesus used it as example of "good measure."

> Give, and it will be given to you. Good measure, pressed down, shaken together, running over, will be put into your lap. For with the measure you use it will be measured back to you.

*Luke 6:38*

J. J. Thomson, the discoverer of the electron, called Reynolds "one of the most original and independent of men," having attended his lectures at Owens College, Manchester. He also described Reynolds' chaotic lecturing style which, though it failed to impart much actual knowledge, "showed the working of a very acute mind grappling with a new problem." His rambling and inconclusive manner of teaching was due in part to his failure to consult the existing literature before developing his own thoughts. He was fond of what Thomson called "out-of-door" physics, including the calming effect of rain or a film of oil on waves, the singing of a kettle and—in the episode that concerns us here—the properties of "sand, shingle, grain and piles of shot." He noted that "ideal rigid particles have been used in almost all attempts to build fundamental dynamical hypotheses of matter," yet it did not appear "that any attempts have been made to investigate the dynamical

---

[3] Jenkin, C. F., 1931, *Proc. R. Soc.*, A **131**, 53–89; Weighardt, K., 1975, *Ann. Rev. Fluid. Mech.* **7**, 89; Dahmane, C. D. and Molodtsofin, Y., 1993, *Powders & Grains*, 93 ed., C. Thornton (Rotterdam: Balkema) p. 369; Horaváth, V. K., Jánosi, I. M., and Vella, P. J., 1996, *Phys. Rev.*, E **54**, 2005–2009.

properties of a medium consisting of smooth hard particles (. . . )," although some of these had "long been known by those who buy and sell corn."

While consistent with his love of out-of-door physics, this preoccupation with sand arose in a more arcane context. Like many others, he had set out to invent an appropriate material structure for the ether of space. The ether was a Holy Grail for the classical physicist. It was also pursued by Lord Kelvin at about the same time, as we shall recount in Chapter 11.

Could the electromagnetic properties of space be somehow akin to the mechanical properties of sand? Reynolds somehow convinced himself of this, asserting that the "ordinary electrical machine" then in use as a generator "resembles in all essential particulars the machines used by seedsmen for separating two kinds of seed, trefoil, and rye grass, which grow together (. . . )."

So inspired was he by this notion that his last paper was entitled "The Submechanics of the Universe." But he hedged his bets by saying that his work also offered "a new field for philosophical and mathematical research quite independent of the ether." Most of his readers probably agreed with J. J. Thomson that this "was the most obscure of his writings, as at this time his mind was beginning to fail." Oliver Lodge diplomatically wrote that "Osborne Reynolds was a genius whose ideas are not to be despised, and until we know more about the ether it is just as well to bear this heroic speculation in mind."

Reynolds' theories baffled his immediate contemporaries and were dismissed by later commentators, especially Sir James Jeans, who could give them "no countenance at all." But among his disjointed ramblings one idea is credited with inspiring others at a critical point in the history of physics. Reynolds thought that by squeezing in some extra grains, a local region of relatively high density could also be created. These regions were to be interpreted as particles in the ether, interacting via its elastic strain. But a region of lesser density could be created, and this would also behave like a particle. These two kinds of particles could annihilate each other. Thus there emerged, in a vague and eventually discredited context, the notion of particles and antiparticles, to be confirmed half a century later.

In his speech at the British Association Meeting (Aberdeen 1885)[4] Reynolds explained that a granular material in a dense state must expand in order to flow or deform

> As the foot presses upon the sand when the falling tide leaves it
> firm, that portion of it immediately surrounding the foot becomes
> momentarily dry (. . . ). The pressure of the foot causes dilatation

---

[4] Reynolds, O., 1885, *British Association Report*, Aberdeen, p. 897; 1885, On the dilatancy of media composed of rigid particles in contact; With experimental illustrations *Phil. Mag.* **20**, 223.

of the sand, and so more water is (drawn) through the interstices of the surrounding sand (...) leaving it dry....

Lord Kelvin spoke admiringly of this observation:

Of all the two hundred thousand million men, women, and children who, from the beginning of the world, have ever walked on wet sand, how many, prior to the British Association Meeting in Aberdeen in 1885, if asked, "Is the sand compressed under your foot?" would have answered otherwise than "Yes?"[5]

What Reynolds observed he called *dilatancy*, since the sand dilates. An expansion is required to allow any deformation (typically density increases of a few percent).

In public lectures he dramatically demonstrated dilatancy (his "paradoxical or anti-sponge property") by filling a bag with sand and showing that if only just enough water was added to fill the interstices, the sealed bag became rigid.

Reynolds was remembered thereafter for his contributions to the dynamics of fluids (including the Reynolds number) but his work on granular materials enjoys belated celebrity today. It has become a fashionable field of physics, one in which fundamental explanations are sought for phenomena long known to engineers.

In one practical example, dry cement is stocked in large hoppers from which it is dispensed at the bottom. Normally, the cement comes out with a constant flux independent of the filling of the hopper (which is not what a normal liquid would do). Occasionally, when it is not been allowed enough time to settle, the cement behaves in a much more fluid manner. A disaster ensues when the hopper is opened.[6]

## 6.3 Major Bagnold's Desert Drive

Fifty years after Reynolds walked on the beach, Ralph Alger Bagnold drove on the desert (Figure 6.2). He too realized that sand has paradoxical properties, that it could exist in different states, and that an appreciation of them could make the difference between life and death when pioneering the exploration of the North African desert. This he undertook, after forays into Sinai and Transjordan, in the vehicles of the 1920s and 1930s. At first this new mode of desert travel was attempted for his own amusement, as

<hr>

[5] Lord Kelvin, 1904, *Baltimore Lectures*, p. 625.

[6] Duran, J., 2000, *Sands, Powders and Grains* (New York: Springer).

**Figure 6.2** R. A. Bagnold, checking his car in the desert. (Copyright from *Libyan Sands*. With permission.)

a junior British army officer. The knowledge and know-how that he accumulated were to serve the Allied cause well into the World War II. His superiors, sipping their drinks in the cool of their club at Gezirah in Cairo, gradually realised the utility of the young chap's folly.

He described his adventures in the book *Libyan Sands*. It was subtitled *Travels in a Dead World*, but Bagnold was intrigued rather than dismayed by that forbiddingly empty world. His immediate preoccupation was survival, and making vehicular progress where none had succeeded before.

Parts of the dunes were "liquid" quicksands, allowing one to "plunge a 6-foot rod vertically into them without effort." Elsewhere there was a surface crust—it all depended on "the special way the grains were packed." Only "something vaguely unfriendly about certain curves and certain faint changes of tint" gave clues where trouble might lie.

To their surprise Bagnold's expeditions found that it was best to drive boldly along the tops of the dunes at high speed, adding greatly to the exhilaration of the trip—at least for a while. Occasionally, the leading vehicle would fail to skate over a patch of quicksand. For this eventuality Bagnold invented devices to get it out. His expeditions were like a one-team Dakar rally, without the helicopters and back-up support.

Not only was the sand beneath his wheels fascinating, but when driven by the wind its grains were very much alive, creating strange dune formations that crept across the landscape, falling lemming-like over cliffs and "rolling inexorably over villages, palm groves and springs."

**Figure 6.3** A sand dune in Namibia.

Later, Bagnold returned to the properties of sand from a more scientific perspective, writing monographs on its properties that set the scene for today's research. Now we understand how the sand particles take off and resettle downwind, creating those oddly shaped dunes and moving them relentlessly forward.[7]

In his conclusion, Bagnold wrote of the romance of discovery, and the centuries-old search for the last, lost oasis of Zerzura, described in Arab legend. He spoke wistfully of the time to come (soon?) when "the experts close their notebooks, for there is nothing else unfound." The world would become "a dull and colorless place."

In the world of research, the chance of stumbling upon Zerzura is never to be discounted. It can be no surprise that the reckless military man turned to the laboratory in search of fresh challenges. His name, perhaps unknown where the desert winds still drive the restless sand, and even Ozymandias was forgotten, is now permanently enshrined in the physical theories of its motion.

## 6.4 Dunes

Dunes are sand hills that have been shaped by the wind, reaching heights up to 500 m (Figure 6.3). Sand is not easy to move and even a strong wind can lift only small grains. So how are the dunes formed? Bagnold

[7] Bagnold, R. A., 1941, *The Physics of Blown Sand and Desert Dunes* (Methuen & Co.: London).

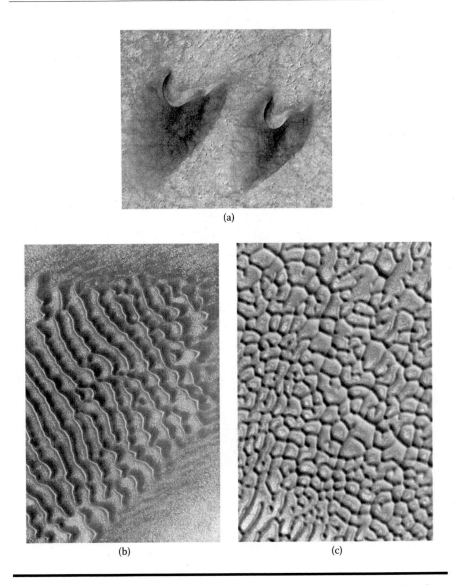

**Figure 6.4** Dunes on Mars: (a) Barcan dunes; (b) transversal dunes; (c) other dune patterns. The Martian dunes are comparable in size to the largest of their terrestrial counterparts. (Mars Orbiter Camera photographs. With permission.)

was the first to study dune formation and he highlighted the mechanism responsible for the movement of large quantity of sand. It is called *saltation*. Small grains are lifted out from the sand bed and accelerated. After a short time, they splash again on the bed ejecting other grains that are in turn caught by the wind. The grains are accumulated into ripples which can turn into ridges and finally dunes. But the same process can move the dunes, as Bagnold observed.

In a typical sandy desert landscape one can observe very gentle hills on the upwind side, where the wind-blown sand is more likely to be deposited, interrupted by sharp edges delimiting regions of steeper downhill slopes. Scientists have classified more than 100 categories of dune shapes, the best known being "longitudinal," "transverse," and "Barchan dunes" (which are crescent shaped, like a new moon) (Figure 6.4).

Dunes can be found in planetary environments where there is a substantial atmosphere, winds, and dust. They are common on Mars, and have been observed on Titan. They provide very useful insights into the nature of the winds on these remote plains.

## 6.5   Order from Shaking

A home-made experiment on spontaneous crystallization can easily be performed in two dimensions by putting small beads on a flat dish and gently shaking it. The result is the triangular arrangement shown in the lower part of Figure 6.5.

Large spheres in a box are not so easily persuaded to form a crystal, but nevertheless shaking them can have interesting effects.

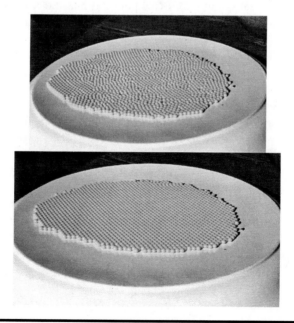

**Figure 6.5**   Spontaneous "crystallization" into the triangular packing induced by the shaking of an initially disordered arrangement of plastic spheres on a dish.

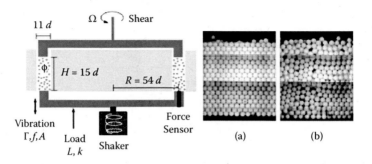

**Figure 6.6** Order in spherical bead packings can be spontaneously induced by simultaneously shaking and shearing.[9] (With permission.)

It was observed in the 1960s that repeated shear or shaking with both vertical and horizontal motion can increase the density of the packing. Recently a packing fraction of 0.67 was obtained by slowly pouring glass beads into a container subject to horizontal vibrations.[8] The resulting structure consists of hexagonal layers stacked randomly one upon the other with a few defects. The order is induced at the boundary and then it propagates vertically layer by layer.

Being the densest (Chapter 3) this packing has the minimum potential energy under gravity. The system spontaneously finds this configuration by exploring the possible arrangements under shaking. The slow pouring lets the system organize itself layer upon layer. This is analogous to what happens in some suspensions where the sedimentation is very slow and the shaking is provided by thermal motion.

In another recent experiment a similar phenomenon has been observed. Shaking combined with shearing can induce beads to crystallize (Figure 6.6) into structures with packing fractions around 0.69.[9] The experiment consists of plastic beads filled into a cell with a spring-loaded bottom to gently squeeze the beads upward, and a rotating top, causing the beads near the brim to move over those below. At the same time, the cell is shaken up and down. What was found was quite counterintuitive: when the shaking was sufficiently vigorous, the flow stopped, and the beads locked into a regular, three-dimensional array like atoms in crystals. In contrast to other physical systems it is as if the system freezes into a crystalline state as temperature is *increased*.

[8] Pouliquen, O., Nicolas, M., and Weidman, P. D., 1997, Crystallization of non-Brownian spheres under horizontal shaking, *Phys. Rev. Lett.* **79**, 3640–3643.

[9] Daniels, K. E. and Behringer, R. P., 2005, *Phys. Rev. Lett.* **94**, 168001.

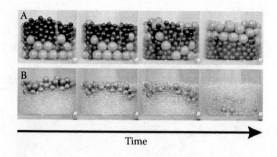

Time

**Figure 6.7** Big particles do not always go to the top. This experiment demonstrates the Brazil nut effect and reverse Brazil nut effect, depending on experimental conditions.[10] (With permission.)

## 6.6   Segregation

Sorting objects of miscellaneous size and weight is a key industrial process. Filtration, sieving or flotation may be used. With granular materials it is often enough to shake the mixture for some time. This must be done judiciously; vigorous shaking (as in a cement mixer) will produce a uniform mixture. Gentle agitation, on the other hand, can promote the segregation of particles of different sizes. The result is quite surprising: the larger, heavier objects tend to rise! This seems an offence against the laws of physics but it is not. That is not to say that it is easily explained. Numerous research papers, including the whimsically entitled "Why the Brazil Nuts Are on Top" have offered theories.[10]

Figure 6.7a shows a simulation of the effect. It is thought to be essentially geometric: whenever a large object rises momentarily, smaller ones can intrude upon the space beneath. Such an explanation calls to mind the manner in which many prehistoric monuments are thought to have been raised. Note that the energy increases as the Brazil nut is raised, contradicting intuition and naive reasoning. This is not forbidden, because energy is being continuously supplied by shaking. Interestingly, under some circumstances the opposite effect can be obtained with small grains at the top and large at the bottom (Figure 6.7b). It has been found that the direction of the segregation depends on the acceleration and frequency of the external shaking, so that it is possible to switch between both effects.[11]

---

[10] Rosato, A., Strandburg, K. J., Prinz, F., and Swendsen, R. H., 1987, Why the Brazil nuts are on top: Size segregation of particular matter by shaking, *Phys. Rev. Lett.* **58**, 1038–1040.

[11] Breu, A. P. J., Ensner, H.-M., Kruelle, C., and Rehberg, I., 2003, Reversing the Brazil-nut effect: Competition between percolation and condensation, *Phys. Rev. Lett.* **90**, 014302.

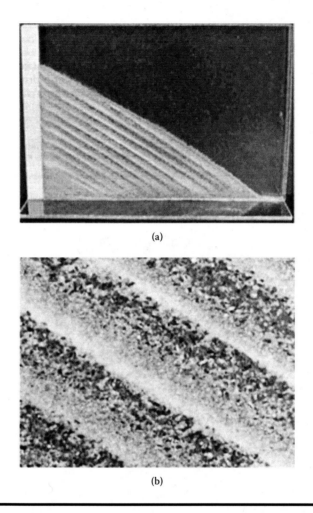

(a)

(b)

**Figure 6.8** Segregation and stratification in a granular mixture of sand, glass beads and sugar poured between two glass plates. (Pictures from Makse, H. A. et al., 1997, *Nature* **386**, 379–382. With permission.)

Separation or stratification into strips can be obtained by slowly pouring a mixture of two kinds of grain of different sizes and shapes between two glass plates (see Figure 6.8). Large grains are more likely to be found near the base of the pile whereas the smaller are more likely to be found near the top. The phenomenon is observable for mixtures of grains in a wide range of size ratios (at least between 1.66 to 6.66 as reported by Makse et al.[12]).

[12] Makse, H. A., Halvin, S., King, P. R., and Stanley, E., 1997, Spontaneous stratification in a granular mixture, *Nature* **386**, 379–382; Fineberg, J., 1997, From Cinderella's dilemma to rock slides, *Nature* **386**, 323–324.

When the large grains have a greater angle of repose[13] with respect to the small grains then the mixture stratifies into strips. (This can be achieved by making the small grains smooth in shape and the large grains more faceted.) For instance, a mixture of white and brown sugar works well for this purpose.

Fineberg has suggested that Cinderella could have utilized this spontaneous separation phenomenon (instead of help from the birds) when the step-sisters threw her lentils into the ashes of the cooking fire.

## 6.7  Granular Temperature

A granular material must have a temperature in the ordinary sense (which we might measure with a thermometer), but this is irrelevant to its behaviour, except in so far as it affects the physical properties of individual particles. The thermal energy is not sufficient to put the grains in motion. However, energy can be provided to the system in other forms, as for instance by continuously shaking the container, as we have described. Given sufficient energy the grains undergo motion, and the system can explore different configurations, analogous to those atoms in thermal systems. This suggests the introduction of a "granular temperature" as a characteristic of such capability to change configuration. Sam Edwards[14] was the first to introduce this notion.[15] Although its many implications are highly debatable (or perhaps because they are so) it has been widely adopted in speculative theories of granular properties.

---

[13] The angle of repose is the maximum angle of a slope in a pile of sand, beyond which it suffers instability in the form of avalanches.

[14] Sir Sam Edwards' achievements have ranged through mathematical physics from elementary particles to the structure of food. As Cavendish Professor he was the successor to four other notable physicists in this book.

[15] Mehta, A. and Edwards, S. F., 1989, *Physica A.* **157**, 1091–1097.

## Chapter 7

# Divide and Conquer: Tiling Space

## 7.1   Packing and Tiling

Packing is an exercise in filling space: tiling (or tessellation) divides it up. The two subjects come together whenever we pack objects that can fill all space, such as the square tiles of the typical bathroom wall.

There are many less banal possibilities than sticking square tiles together. They were explored and exploited by the early Muslim culture that left such masterpieces as the Alhambra for us to admire. Today's mathematicians study a vast range of tiling patterns that echo that preoccupation with regularly repeating patterns.[1] Such a pursuit can be extended to three dimensions, although its applications are less familiar.

Tiling patterns can be constructed by hand, by trial and error, or generated mathematically. One way of doing this is particularly simple and significant. In two dimensions, let us arrange an array of points in a plane, and make these the centers of the tiles (or cells), by including in each tile all points closer to its center than any other. We will call this the Voronoï construction.

---

[1] Grunbaum, B. and Sheppard, G. C., 1987, *Tilings and Patterns* (W. H. Freeman: New York).

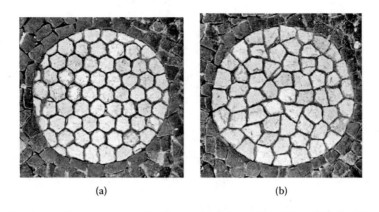

(a)                                                    (b)

**Figure 7.1**   Ordered and disordered tessellations, from the pavements of Lisbon.

## 7.2   The Voronoï Construction

The Voronoï construction is a geometrical construction which recurs throughout this book. It has become attached to the name of Voronoï (1868–1908) but the primary credit probably belongs to Dirichlet, in 1850. Both he and Voronoï used the construction for a rather abstract mathematical purpose in the study of quadratic forms. Since then, it has played a supporting role in many important theories. It consists in constructing around the center of each packed object a *cell* which contains all the points closest to such center than any other center in the packing. The set of all Voronoï cells is a space-filling cellular partition.

The Voronoï construction (Figure 7.2) gives a definite meaning to "near neighbors" for a given set of points. We may count as near neighbors only those pairs of points whose cells are in contact. It is also a device for constructing geometrical proofs of the kind described in this book (see Chapter 4 and Appendix A).

The mathematician Georgy Fedoseevich Voronoï (or Voronoy) (Figure 7.3a) was born in 1868 in the village of Zhuravki in Ukrainia, but from 1889 he studied at St. Petersburg University. As member of the St. Petersburg school of mathematics he made a number of important contributions, but published only 12 papers, hardly enough to survive in the present academic world. Much of his work is still celebrated, but none of it is as widely applied as his division of a plane into a pattern of Voronoï polygons. Mathematically inclined scientists of every kind have found the Voronoï construction useful. So wide are its applications that a variety of

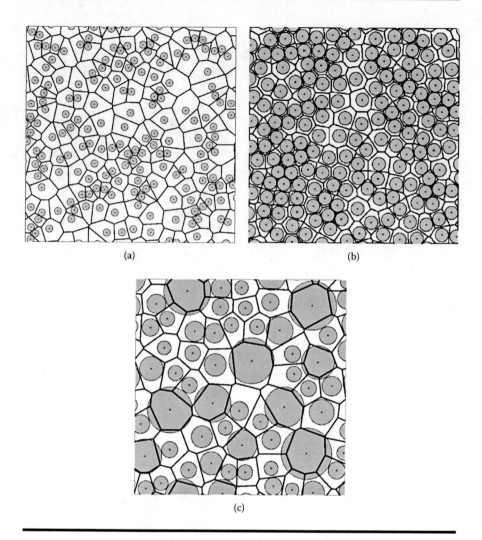

**Figure 7.2** (a, b) Voronoï patterns from disordered disk arrangements. (c) When the disks have different sizes the Voronoï cells no longer contain the entire disk. A generalization that overcomes this problem consists in building the cell boundaries at an equal distance from the borders of the packed objects instead of at an equal distance from their centers.

nomenclature has arisen, including Thiessen polygons (for n-dimensional constructions), Wigner–Seitz cells (for atomic systems), Wirkungsbereich (domain of influence), and Dirichlet cells: Dirichlet developed the concept half a century before Voronoï, but in science it is often the runner in second place that is awarded the gold medal.

(a)                    (b)

**Figure 7.3**    (a) Georgy Fedoseevich Voronoï (1868–1908) and (b) Boris Nikolaevich Delaunay (1890–1980).

## 7.3    The Dual Construction of Delaunay

Once Voronoï has cut up space in his celebrated style, we may join up centers of adjacent cells to make a dual pattern associated with the name of Boris Nikolaevich Delaunay (Figure 7.3b), or Delone, in a form closer to the Russian name. Not everyone realises that this is the same person. His surname comes from his Irish ancestor Deloney, a mercenary in Russia with Napoleon. Delaunay is the French transliteration of the name.

## 7.4    Applications

The Voronoï construction has many applications in computer science and mathematics, but its broad appeal is more associated with its interpretation as an "area of influence." The modern use of Voronoï diagrams in physical science (in two, three, or in many dimensions) began with crystallography but has since become much more general. In geography and ecology, indeed everywhere that spatial patterns are analyzed, this construction has proved useful. The German term "Wirkungsbereich" for a Voronoï cell is particularly apt; it refers to a region of activity or influence. It has often been considered to be a good basis for dividing political territories, given a set of preexisting centers in major towns. Indeed, the division of France into Departments, laid down by Napoleon, corresponds rather well to the

Voronoï regions around their principal cities. How appropriate in the home of rationality, a country that likes to call itself "l'Hexagone!"

Territories that arise more naturally by outward growth from centers, such as plant colonies, may be expected to resemble Voronoï patterns, and they often do. Voronoï constructions are useful in both two and three (and higher) dimensiions. Even metallurgists look for inspiration here, in analysing the arrangements of the crystalline grains that are formed by re-crystallisation, starting from random centers. In solid state physics the name Wigner–Seitz cell has been used instead, since 1933. This was used to calculate the changes in the energy of electrons when atoms were packed together.

A recent monograph on the subject, although confined largely to two-dimensional patterns, ran to more than 500 pages, so it evidently has many ramifications.[2]

## 7.5   Vertices in Tilings

The examples of Voronoï tilings that we have already admired have several features in common, which the reader may suspect are general. First, the Voronoï polygons are convex (they do not point inwards at any vertex) and have straight edges. This is indeed always the case. Second, the vertices formed by their corners are all threefold in these examples. This is not entirely general. For example, a square array of centers will produce the square tiling pattern, with fourfold vertices. But special symmetries are required for this. All patterns that have any degree of randomness will conform to the rule that only threefold vertices can occur. This implies that the Delaunay construction cuts the plane up into triangles.

With a few exceptions that we leave the reader to find, national state boundaries conform to the same rule, though they may be far from Voronoï constructions.

## 7.6   Three Dimensions

The Voronoï construction in three dimensions generates polyhedra with flat sides (Figure 7.4), again convex. In the general case, *three* polyhedra are joined at each edge, and *four* at every vertex, and the Delaunay dual pattern consists of tetrahedra. These tetrahedra pack together to fill all space, which may seem to contradict our earlier declaration of the

---

[2] Okabe, A., Boots, B., and Sugihara, K., 1992, *Spatial Tessellations: Concepts and Applications of Voronoï Diagrams* (Chichester: Wiley).

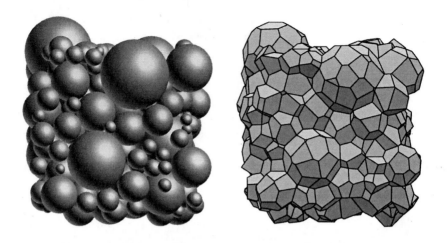

**Figure 7.4** Obtained from two-packing of equal spheres (left) and three-dimensional Voronoï partition (right). (Courtesy of Andrew Kraynik.)

impossibility of such packing. But that assertion was made for regular tetrahedra, and the Delaunay tetrahedra are not so. We can fill space with tetrahedra, in just this way, if we do not insist on their regularity.

Many ordered array of points in three dimensions also give rise to Voronoï/Delaunay patterns as we have described. The simplest example is the body-centered-cubic lattice (Figure 8.5c).

## 7.7 Regular and Semiregular Packings

We can now associate the triangular packing of equal disks with the hexagonal honeycomb structure of cells that arises if the discs are used in a Voronoï construction (or tessellation). The dual Delaunay tessellation is made of equal, regular triangular tiles that join the disk centers.

This particular tessellation has three properties:

(1)   all the vertices are identical, that is, lines come together in the same way at each of them;
(2)   all tiles are regular (that is, completely symmetrical) polygons;
(3)   all polygons are identical.

It is therefore called a *regular tessellation.* How many tessellations with such regularity exist? The answer is three: they correspond to the *triangular, square* and *hexagonal* cases, which are shown in Figure 7.5. In these three

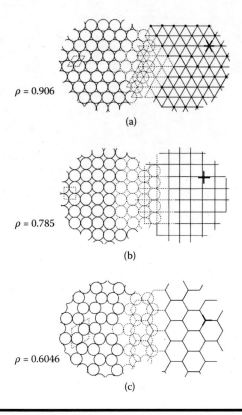

$\rho = 0.906$

(a)

$\rho = 0.785$

(b)

$\rho = 0.6046$

(c)

**Figure 7.5** Regular packings of equal discs. Here the patterns formed by joining the centers of the contacting circles have identical vertices and regular, identical polygons as tiles. These are triangles (a), squares (b), and hexagons (c). The packing fractions are respectively $\rho = \pi/\sqrt{12} = 0.906\ldots$ (a), $\rho = \pi/4 = 0.785\ldots$ (b), and $\rho = \pi/\sqrt{27} = 0.604\ldots$ (c).

packings the discs are locally disposed in highly symmetrical arrangements and the whole packing can be generated by translating on the plane a unique local configuration, as in the simplest kinds of wallpaper.

By relaxing the third condition, and allowing more than one type of regular polygon as tiles, eight other tesselations can be constructed. These tessellations are named *semiregular* or *Archimedean* by analogy with the names used for finite polyhedra in three dimensions. Figure 7.6 shows these eight semiregular packings and their packing fractions. Note that the lowest packing fraction among these cases ($\rho \simeq 0.391$) is attained by the structure made with triangular and dodecagonal tiles.

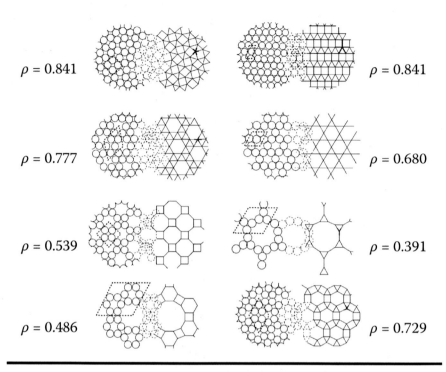

ρ = 0.841             ρ = 0.841

ρ = 0.777             ρ = 0.680

ρ = 0.539             ρ = 0.391

ρ = 0.486             ρ = 0.729

**Figure 7.6** Semiregular packings of equal discs. Here the patterns formed by joining the centers of the circles mutually in contact have identical vertices and tiles, which are different kinds of regular polygons.

In these semiregular packings the whole structure can again be generated by translating a unique local configuration. But nature is rich and diversified, and several other types of packings different from the simple crystalline ones are also found.

When discs with special size ratios are chosen, beautiful ordered arrangements can also be created. This, for instance, is the case when equal quantities of discs with two diameters in the ratio $\sqrt{(3-\tau)/\tau}$ are chosen (here, $\tau = (1 + \sqrt{5})/2 = 1.618\ldots$, is the "golden ratio," which is the ratio between the base and the height of the *golden rectangle*: the shape with perfect proportion for the ancient Greeks). In Figure 7.7 different arrangements obtained by using this mixture are shown. (The diameter ratio of the U.S. cent and quarter coins is very close to this ratio $\tau$, and their contrasting colors make nice figures.)

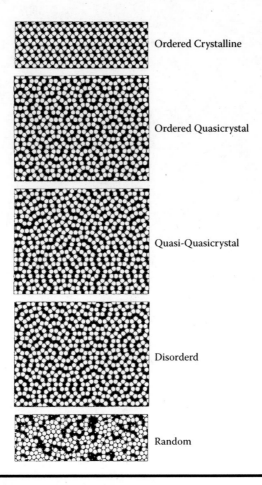

Ordered Crystalline

Ordered Quasicrystal

Quasi-Quasicrystal

Disorderd

Random

**Figure 7.7** Various tilings obtained using a binary mixture of discs. (From Lançon, F. and Billard, L., 1995, Binary tilings tools for models, *Lectures on Quasicrystals*, eds., F. Hippert and D. Gratins (*Les Vlis: Ed. de Physique*) pp. 265–281. With permission.)

# Chapter 8

---

# Peas and Pips

---

## 8.1   Vegetable Staticks

When soft objects are tightly packed, they change their shapes to eliminate the wasted interstitial space. Even if they begin as spheres they will develop into polyhedra. The question is: *Which types of polyhedra will be formed?* The formation of foam by bubbles is an example of such a process; another is to be found in the familiar insulating material made of polystyrene, formed by causing small spheres to expand and fill a mold. But the most famous early experiment was carried out with peas.

The Reverend Stephen Hales performed this classic experiment, in an age when science was practiced as much in the parlor or the kitchen as in the laboratory. Hales compressed a large quantity of peas (or rather expanded them under pressure by absorption of water) and described what he observed in a book with the charming title of *Vegetable Staticks*.

> I compressed several fresh parcels of Pease in the same Pot, with a force equal to 1600, 800, and 400 pounds; in which Experiments, tho' the Pease dilated, yet they did not raise the lever, because what they increased in bulk was, by great incumbent weight, pressed into the interstices of the Pease, which they adequately filled up, being thereby formed into pretty regular Dodecahedrons.[1] (Figure 8.1)

---

[1] Published in 1727 by Stephen Hales, with the title *Vegetable Staticks*: Or *An Account of Some Statistical Experiments on the Sap Vegetables*, being an essay towards a "Natural History of Vegetation." Also, "A Specimen of an Attempt to Analyse the Air, by Great Variety of Chymio-Statical Experiments," which were read at several meetings before the Royal Society and dedicated to His Royal Highness George Prince of Wales.

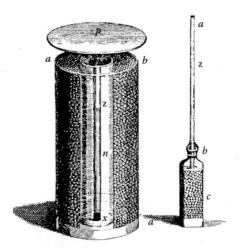

**Figure 8.1** The experimental apparatus used by Hales to demonstrate the force exerted by dilating peas. When the lid was loaded with a weight, the dilated peas filled the interstices, developing polyhedral forms.

At this point the Reverend gentleman's report is misleading. We will see later in this chapter that when soft objects are closely packed in a disordered fashion, they form polyhedral grains with irregular shapes. Only a few of them have 12 faces. Moreover, regular dodecahedral cells[2] cannot fill three-dimensional space, so Hales purported to observe something which is plainly impossible. But this mistake probably has a simple explanation: in such packings the majority of faces are pentagons and the dodecahedron is the regular solid made with pentagonal faces. The history of science is full of such cases in which the observer tries to draw a neat conclusion from complex and variable data. Too determined a search for a simple conclusion can lead to an erroneously idealized one.

Rob Kusner (a mathematician at the University of Massachusetts at Amherst) tells us that a modern version of Hales' experiment with peas was carried out in New England a few summers ago by a group of undergraduates using water-balloons, greased with vegetable oil and stuffed into a large chest freezer. Some balloons burst from being pierced by sharp ice crystals, and others did not freeze, but it was still possible to see the packing patterns since the low temperature folding on the balloon rubber left permanent marking (lighter color) at the folds.

Today, this experiment can be done by means of x-ray tomography (Section 5.6). Figure 8.2 shows the resulting image with the parts under stronger compression highlighted in darker tones.

---

[2] The reader unfamiliar with their polyhedra should turn to Figure 8.5.

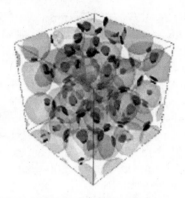

**Figure 8.2** Tomographic image of rubber balls undergoing compression. (Courtesy of M. Saadatfar.)

## 8.2 Stephen Hales

At first sight, the pea-packing experiments of Stephen Hales (Figure 8.3) might seem to be the dabblings of a dilettante. Not so. Hales was a significant figure in the rapid development of science after Newton. He is much mentioned, for example, in the work of Lavoisier, and has been the subject of several biographies and many portraits.

Belonging to the gentry of the south of England, he had no difficulty in gaining access to undemanding positions as a clergyman, many of which were in the gift of members of his class. The research that he carried out over many years forms part of the foundation of today's science of the physiology of plants and animals—the study of function as opposed to mere form, as in anatomy and taxonomy. He was also something of a

**Figure 8.3** Stephen Hales (1677–1761).

technologist, being credited with the invention of forced ventilation. It has not always been self-evident that fresh air is good for us! The Navy, in particular, enthusiastically adopted his recommendations in an attempt to improve the health of sailors.

It was in a physiological spirit that Hales performed his experiment with dried peas. The pressure associated with their uptake of water was at issue: what we would today call "osmotic pressure." The fact that the peas, swelling while under pressure, were compressed into polyhedral forms, was really incidental. This may be offered as an excuse for the unfortunate inaccuracy of Hales' description of their polyhedral form.

It would be pleasing to establish some family connection between this man and his modern namesake (Chapter 4). Alas, Thomas Hales reports that none has been established, and reminds us that Stephen Hales died without issue.

## 8.3   Pomegranate Pips

About one century before Hales, Kepler had studied the shape of the pips inside a pomegranate (Figure 8.4). These seeds are soft juicy grains with polyhedral shapes. Kepler was trying to understand the origin of the hexagonal shapes of snowflakes, so the pomegranate grains were used as an example of the spontaneous formation of regular geometrical polyhedral shapes in a packing. Kepler observed that these grains have rhomboidal faces and he remembered that the same rhomboidal faces are present in the bottom of bees' cells and make up the interface between the two opposite layers of cells in the honeycomb (see Chapter 9).

**Figure 8.4**   A pomegranate.

Attracted by these rhombes, I started to search in the geometry if a body similar to the five regular solids and to the fourteen Archimedean could have been constituted uniquely of rhombes. I found two (...). The first is constituted of twelve rhombes, (...) this geometrical figure, the closest possible to the regularity, fill the solid space, as the hexagon, the square and the triangle fill the plane.

(...) if one opens a rather large-sized pomegranate, one will see most of its loculi squeezed into the same shape, unless they are impeded by the peduncles that take food to them.

(...) What agent creates the rhomboid shape in the cells of the honeycomb and in the loculi of the pomegranate?

Kepler argues that the key to the honeycomb and pomegranate is the problem of packing equal-sized spherical objects into the smallest possible space; and he finds the answer in a conjecture for the closest sphere packing (Chapter 3). Indeed, if one takes this packing and expands the spheres, grains with 12 rhombic faces are obtained.

Kepler carefully warns the readers that one must take a "rather large-sized pomegranate" to find many rhombohedral grains inside. Indeed, in a typical pomegranate the rhombohedral grain shape is not so common. The packing is not so perfect and the grains take different shapes. Just like Stephen Hales, Kepler was oversimplifying his conclusions, or "keplering" (Chapter 3).

## 8.4  Biological Cells, Lead Shot, Rubber Balls, and Soap Bubbles: *Plus ça Change*

From the beginning of microscopy, anatomists were impressed by the similarities between the shape of biological cells in undifferentiated tissues and that of bubbles in foam. Robert Hooke includes in his *Micrographia* (1665) an observation on the "Schematism or Texture of Cork, and the Cells and Pores of Some Other Such Frothy Bodies" and describes the pith of a plant as "congeries of very small bubbles." For centuries, with almost no exceptions, cells in undifferentiated tissues were described as regular dodecahedra (Figure 8.5a) (the impossible peas of Hales) or as rhombic dodecahedra (Figure 8.5b) (the improbable pomegranate seed of Kepler). Then, at the end of the nineteenth century a new type of cell with 14 faces was promoted. It was the *tetrakaidecahedron* (Figure 8.5c), a polyhedron that Lord Kelvin proposed in 1887 as the structure that divides space "with minimum partitional area." (We will follow Kelvin's line of thought in Chapter 11.)

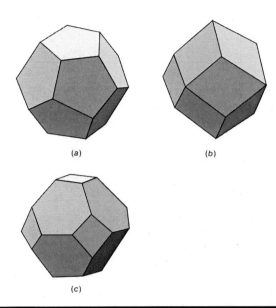

**Figure 8.5** A pentagonal regular dodecahedron (a), a rhombic dodecahedron (b), and a tetrakaidecahedron (c).

It was only at the beginning of this century that careful and extensive studies of the shape of bubbles in foams and cells in tissues were undertaken. In particular, a large series of biological tissues was meticulously studied by F. T. Lewis of Harvard University between 1923 and 1950.[3] He concluded that cells in undifferentiated tissues have polyhedral shapes with about 14 faces on average and he inferred that they tend to be approximated by Kelvin's ideal polyhedron. Further investigation now indicates that the cells have about 14 faces on average but a large variety of shapes contribute to form the cellular structure. The Kelvin polyhedron is rarely observed.[4]

The experimental study of the form of biological cells inspired several experiments in which cellular structures were created by compressing together soft spheres to fill all the space (as Reverend Hales did with peas). In this way, the resulting structure can be disassembled and the shapes of the individual cells easily studied. A classical experiment of this kind

[3] See, for example, Lewis, F. T., 1950, Reciprocal cell division in epidermal and subepidermal cells, *Am. J. Bot.* **37**, 715–721.

[4] Dormer, K. J., 1980, *Fundamental Tissue Geometry for Biologists* (Cambridge: Cambridge University Press).

is the one by Marvin,[5] who compressed 730 pellets of lead shot in a steel cylinder. When the spheres are carefully packed layer by layer in the closest way, then—not surprisingly—the cells take Kepler's rhombic dodecahedral shape. Totally different shapes are observed when the spheres are packed in a disordered way, for instance by pouring the shot in the container or by shaking it before compression. In this case, the cellular structures have polyhedral cells with faces that vary from triangles to octagons with the great majority being pentagons and, less abundantly, squares and hexagons. The average number of neighbors was reported to be 14.17 in a set of 624 internal lead pellets. No Kepler cells were observed in these experiments.

One may look for similar structures in deformed bubbles of foam, but looking at bubbles inside a foam is quite difficult, if we wish to observe the properties of *internal* bubbles. In 1946, Matzke[6] reported the study of the shapes of 600 central cells (which he claimed to be of equal volume) in a soap froth. This research is still the most extensive experimental study of the structure of foam bubbles ever undertaken. He found an average number of faces per bubble of 13.7 and a predominance of pentagonal faces, with 99.6% of all faces being either quadrilateral, pentagonal or hexagonal. Kepler's cell never appeared, nor was that of Kelvin observed. This tale is told in more detail in Chapter 11.

Figure 8.2 shows a compressed packing of rubber balls. In many respects it resembles the packing of bubbles that we call a foam (Chapter 11). Under sufficient compression the balls will be reduced to polyhedral shapes, with no space between grains.

Biological cells, peas, lead shot and soap bubbles are quite different systems, although all consist of polyhedral cells packed together to fill the whole space. A biological tissue is generated by growth and mitosis (division) of cells. The shape of a cell is therefore continuously changing following the mitotic cycle. In contrast to this, when lead shot is compressed the shapes of the cells are mostly determined by the environment of the packed spheres. Indeed, during the compression, rearrangements are very rare. In foams the structure is strictly related to the interfacial energy and bubbles assume shapes that minimize the global surface area. The great similarity in the polyhedral shapes of the cells in these very different systems can therefore be attributed only to the inescapable geometrical condition of filling space.

---

[5] Marvin, J. W., 1939, The shape of compressed lead shot and its relation to cell shape, *Am. J. Bot.* **26**, 280–288.

[6] Matzke, E. B., 1946, The three-dimensional shape of bubbles in foam—An analysis of the role of surface forces in three-dimensional cell shape determination, *Am. J. Bot.* **33**, 58–80.

## Chapter 9

# Enthusiastic Admiration: The Honeycomb

## 9.1 The Honeycomb Problem

We have encountered various cases of *cellular structures*, which divide space into cells. How can this be done most economically, in terms of the surface area of the cells? It is not clear that this has any relevance to the squashed peas of Hales or the lead shot of Marvin, but it certainly is the guiding principle for foams (the subject of Chapter 11), for which the cell interfaces cost energy. The bubble packing which we call a foam is not alone in minimizing surface area. Emulsions, such as that of oil and vinegar shaken to make a salad dressing, conform to the same principle.

For centuries this principle has also been supposed to govern the construction of the honeycomb by the bee. The bee, it has been said, needs to make an array of equal cells in two dimensions, using a minimum of wax, and hence requires a pattern with the *minimum perimeter per cell*.

Although the perfection of the honeycomb is a very proper object for admiration, it may be naive to impute to the bee the single mathematical motive of saving of wax, just as it can hardly be said that the greengrocer cares much about the maximum density of oranges. There really aren't many reasonable alternatives to the two-dimensional hexagonal structure for the honeycomb. Other considerations surely impose themselves, such as simplicity and mechanical stability, in the evolutionary optimization of the hive.

A full account of the arguments that have raged over the shape of the bee's cell would read like a history of Western thought. We can find one of the first attempts at an explanation in Pliny (*Naturalis Historia*) who associated the hexagonal shape of the cell with the fact that bees have six legs. Among the other notable minds that have been brought to bear on it, we must count at least those of: Pappus of Alexandria (*Fifth Book*), Buffon (*Histoire Naturelle*), Kepler, Koenig, Maraldi, Réamur, Lord Brougham, Maurice Maeterlinck (*La Vie des Abeilles*), Samuel Haughton, Colin MacLaurin, Jules Michelet (*L'insecte*), and Charles Darwin (*The Origin of Species*). "He must be a dull man," said Darwin, who could contemplate this subject "without enthusiastic admiration." Not surprisingly, he attributed "the most wonderful of all known instincts" to "numerous, successive, slight modifications of simpler instincts."

Darwin's account of the process by which the honeybee achieves its precise constructions, by forming rough walls and refining them, is instructive, but he is not quite correct in saying that "they are absolutely perfect in economizing wax" as we shall shortly see, when we turn to the three-dimensional aspect of the hive, that is, the structure of the interface between the two opposed honeycombs. For the moment, we address only its two-dimensional aspect, the arrangement of the elongated cells which is visible on the surface. Does this two-dimensional pattern of cells of equal area have the least possible perimeter?

Of course it does: it is well known that this pattern is the best. What has remained hidden from general appreciation is that this proposition has not until now been fully proved! This was rarely stated, probably because most authors could not believe there is no proof concealed somewhere in the unfathomable depths of the technical literature. We saw in Chapter 2 that a proof exists for the closely related problem of optimal packing of equal discs, but this should not be confused with the question posed by the honeybee.

This should take its rightful place alongside Kepler's problem as a notable frustration for the mathematician.[1]

There did exist a proof of a lesser theorem, once again attributed to L. Fejes Tóth (Figure 9.1). It imposes certain restrictions, of which the most important is the requirement that all the sides of the cells are straight. This follows from the convexity principle which was described in Chapter 2. But in general it is very natural for them to be curved, so this is a much weaker result than one would like.

At the time of our first edition, Thomas Hales informally announced that he would shortly publish such a full proof.[2] This and his proof of the Kepler Conjecture constitute a remarkable double achievement in a short space of time.

---

[1] Morgan, F., 1999, The hexagonal honeycomb conjecture, *Trans. Am. Math. Soc.* **351**, 1733; 2000, *Geometric Measure Theory: A Beginners Guide*, 3rd Ed. (New York: Academic).

[2] http://www.math.pitt.edu/articles/cannonOverview.html.

**Figure 9.1**   The Hungarian mathematician László Fejes Tóth was a leader in the mathematics of packings for many years, and his son Gábor Fejes Tóth now follows in his footsteps.

```
Date: Mon, 7 Jun 1999 12:37:27 -0400 (EDT)
From: Tom Hales
To: Denis Weaire
cc:
Subject: Honeycombs
```

Dear Denis Weaire,

If all goes well, I'll announce a solution to the honeycomb conjecture in a few days.  (I don't make any assumptions about the convexity or topology of the cells.)  I've shown it to Frank Morgan and John Sullivan, and they didn't find any obvious problems with my proof.

I hope you don't mind that in my acknowledgements, I quote from your email message to me from October, ``Given its celebrated history, it seems worth a try...'' Thanks for attracting my attention to the problem.

Of course, I'm fascinated by the Kelvin problem too, but I don't think that will be solved anytime soon...

Best,
       Tom

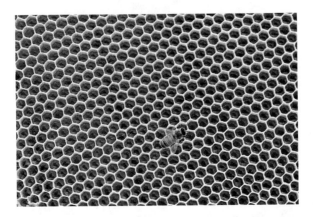

**Figure 9.2** The beehive.

The proof[3] is much less intimidating than that of the Kepler conjecture. The reaction of many colleagues must have flickered between a smile of admiration and a grimace of disappointment, that this had escaped their grasp.

## 9.2 What the Bees Do Not Know

The enthusiastic admiration of the scientific community for the pattern of the beehive (Figure 9.2) has often extended to the *internal* structure of the honeycomb. It has two opposite sides from which access is possible, with a partition wall in the middle. But in this case, as we have warned, it turns out that bees (even Hungarian ones) are not such punctilious mathematicians after all, when the three-dimensional aspect of their construction is considered.

A flat wall would be wasteful of wax. Instead, the bee chooses a faceted wall which neatly fits the two halves of the honeycomb, when the cells are staggered with respect to each other. The edges of the facets meet each other and the side walls of the hexagonal cells at the same angle (the Maraldi angle) of approximately 109° (see Figure 9.3b). As we will see in the next chapter, Plateau recognized this in foams as a consequence of the minimization of surface energy. So, the parsimony of the bee apparently extends even to this internal structure. (Today, the bee has no real choice, since the wall is provided by the keeper as a preformed "foundation.")

There is a close connection between this strategy and the successive optimal stacking of close-packed planes of spheres. The Voronoï construction (Section 7.2) applied to two such layers gives, for the partition between them, precisely the form of the bee's wall.

---

[3] Hales, T. C., 2001, The Honeycomb Conjecture, *Discr. Comput. Geom.* **25**, 1.

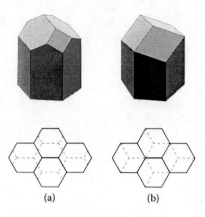

(a)  (b)

**Figure 9.3**   The alternative of Fejes Tóth (a) to the bee's design (b).

Indeed, the bee's faceted wall can be shown to minimize area and hence the expenditure of wax, in the limited sense that any small change will increase the area. The occurrence of the Maraldi angle signals this: indeed it gained its name in this context.

Learned academies have sung the praises of the bees for basing their construction on the Maraldi angle and, in so doing, have exaggerated its precision. This has caused some speculation as to whether the Almighty has endowed these creatures with an understanding of advanced mathematics.

The story is told in detail in D'A. W. Thompson's *On Growth and Form*.[4] Dismissing what he considered to be earlier follies (including Darwin's) he is driven to the extreme conclusion that "the bee makes no economies; and whatever economies lie in the theoretical construction, the bee's handiwork is not fine or accurate enough to take advantage of them." Where Darwin had invoked slow changes in response to a marginal advantage of design, Thompson looked for physical forces at work, supposing the thin wax film of the hive to be more or less fluid at the same time in their construction, and so forming the angles dictated by surface tension.

Where precisely the truth lies in this old and muddled dispute about angles we do not know, but Thompson would clearly have been delighted to learn of L. Fejes Tóth's startling conclusion, many years later: there is an entirely different arrangement which is better than the design of the bee! This was published in a charming paper entitled "What the Bees Know and What They Do Not Know."[5]

[4] Thomson, D'A. W., 1942, *On Growth and Form*, 2nd Ed. (Cambridge: Cambridge University Press).

[5] Fejes, Tóth L., 1964, *Bull. Am. Math. Soc.*, **70**, 468–481.

The bee's design can be improved, with a saving of 0.4% of the surface area of the wall, by using a different arrangement of facets, again constructed with the angles dictated by surface tension. The alternative presented by L. Fejes Tóth is shown in Figure 9.3 and is closely related to the Kelvin structure to be described in Chapter 11.

Bees do sometimes create the Hungarian mathematician's design locally, whenever they are left to build the wall themselves, and then make a mistake, so that the two parts of the honeycomb are misaligned. But otherwise they seem to opt for simplicity.

# Chapter 10

# A Search for Structure

## 10.1 A Voice in the Wilderness

In the 1950s, Cyril Stanley Smith (Figure 10.1), a metallurgist who had played a leading part in the Manhattan project, engaged in what he called *A Search for Structure*, when he summarized his second career in a book. He had become an incomparable historian of the science and technology of materials. He also tried to formulate general ideas of form and organisation, to supplement or replace more traditional ones. Not all of nature, he insisted, is ordered, smooth, continuous. Much of it is disordered, rough and ragged, jagged and jerky. "Scientists have often overlooked the [true] form of the world."

**Figure 10.1**  Cyril Stanley Smith (1903–1992).

His lone quest for grand new concepts did not arrive at a coherent outcome, but it did find concrete expression in one piece of work which bridged the two phases of his professional life. It was based on his conception of the two dimensional soap froth as a prototype for metallic grain growth (and much else). He rightly sensed that it would be an enduring source of understanding of some of the complex effects that he so admired in materials, and that these had general import.

In particular he insisted:

> A complex structure is the result of and to a large extent the record of its past.

He coined a word for this property of dependence on history. It was to be *funicity* (hence *funeous* and *afuneous*), after a character in a short story of Borges, who sadly lost his capacity to forget things. It deserves to be used. So let us not forget to do so.

He also stressed the importance of nucleation of change and growth:

> If you can look into the seeds of time and say which grains will grow and which will not, speak then unto me.

He thought that scientists should be unashamed of a sense of fun and a propensity to play. It must have been in that spirit that he first made for himself a two-dimensional soap froth, as a model counterpart of metallic grain structure.

## 10.2   The Two-Dimensional Soap Froth

Even in the face of great complexity (and metallurgy is a mass of detail) we must try to simplify, to strip out extraneous or incidental complications and retain the essence in an idealised model and/or prototypical physical system. That is part of the scientific method.

Smith advanced the two dimensional soap froth (Figures 10.2 and 10.3) as the best representative of many cellular systems controlled by surface energy (or tension). Its merits include its two-dimensional nature, which was to facilitate measurement and analysis and eventually also simulation by computer.

Made by squeezing an ordinary soap froth between two glass plates, Smith's packed, flattened bubbles formed polygonal cells akin to the grains of a polycrystalline metal, and they behaved similarly, especially in regard

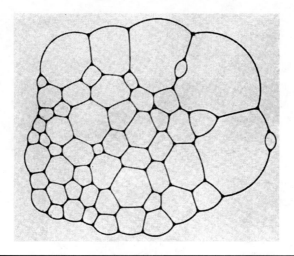

**Figure 10.2** Two-dimensional soap froth. (From Smith, C., 1964, Structure, substructure, and superstructure, *Rev. Mod. Phys.* **36**, 524–532. With permission.)

to grain growth, as gas slowly diffused between the bubbles. This is called coarsening.

In recent decades the physics, chemistry and engineering of foams has developed a coherent community of adherents, reviving a venerable topic that attracted many great names in the distant past. For this community, whose figurehead has been Pierre-Gilles de Gennes,[1] Nobel Prizewinner for contributions to the physics of soft matter, Smith's two dimensional system is no longer merely a simplistic stand-in for a polycrystalline solid. It is the best starting point for the admiration and analysis of foam structure in general, and other forms of soft matter.

Even more expansively, Smith saw his model system as something universal. The minimization of energy, the elementary rearrangements that adjusted the structure, its evolution and destruction: all seemed to be prototypical of the natural world. This universality is indeed appealing: at very least the life and death of the soap froth offer a powerful metaphor, as poets have already recognized.

## 10.3 The Rules of the Game

Ordinary two-dimensional foam is quite disordered, but close examination reveals local rules of equilibrium to which it must conform. This is one corner of the subject that was developed in full by Plateau, the hero of

---

[1] He died in 2007 while we were writing these pages.

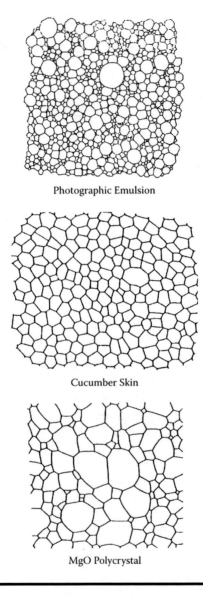

Photographic Emulsion

Cucumber Skin

MgO Polycrystal

**Figure 10.3** Two-dimensional cellular patterns. (From Weaire, D. and Rivier, N., 1984, Soap, cells, and statistics—Random patterns in two dimensions, *Contemp. Phys.* **25**, 59. With permission.)

our next chapter. For the moment, note that all the vertices of the pattern involve no more than three lines, and those lines are always circular. Closer scrutiny reveals that their curvatures are governed by the pressure differences between the cells. The eye senses and appreciates this inner order, the emergent effect of unseen forces.

But what of the overall structure, which, as Smith insisted, is funeous. The same bubbles can be rearranged in countless ways, in terms not just of their individual placement, but even the overall statistics of the structure. No grand principle of thermodynamics is at hand to dictate what kind of structure it must be—for example, with small bubbles segregated.

Smith insisted that this need not pose a total block on intelligent analysis of the structure. It does exhibit reproducible statistical properties, and people still argue about them. Furthermore, there does exist a procedure that entails a unique well-defined structure, not some arbitrary hodge-podge. It is simply to let the process of coarsening have its way for a little while. Magically, as the structure increases in scale due to coarsening (as cells get eaten up by neighbors), it settles down into a certain scaling state. Thereafter, cells continue to disappear and be shuffled around, but statistically speaking it does not change.

All of which and much more may be studied in the same playful manner as Smith, displayed on an overhead projector, recorded on a photocopier. This is the extreme opposite end of science from that of Superconducting Supercolliders. Without disparaging the latter, there may be a message here for the taxpayer.

## 10.4   In a Cambridge Garden

Although more conventional than Smith, his contemporary William Lawrence Bragg (Figure 10.4) shared the same wonder at the forms of nature. Having been awarded a Nobel Prize for experimental crystallography (Chapter 13),

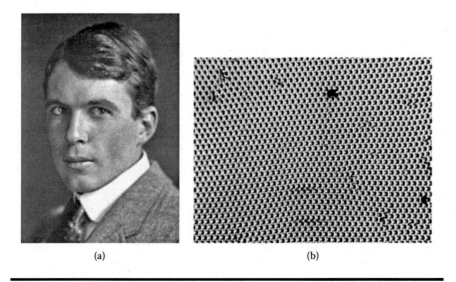

(a)                                    (b)

**Figure 10.4**   (a) William Lawrence Bragg (1890–1971) and (b) his bubble raft.

he may be excused for a preoccupation with perfectly ordered structures, but the messier aspects of metals engaged his attention as well. Why are they not as strong as simple theory leads us to expect? What defects are at work? He arrived at similar problems to those of Smith from another direction, that of the physicist, and he saw the same need for a model system.

Bragg steered the University of Cambridge away from nuclear physics towards the experimental pursuit of the arrangement of atoms, which was eventually to pay off spectacularly in the revelation of the structure of DNA by Crick and Watson.

In the mid 1940s John Nye joined him as a research student, and began to adjust to the intimidating atmosphere of the Cavendish Laboratory, where even a shoddy piece of electrical cable was grudgingly dispensed with the admonition: "Rutherford used it, bring it back!"

Bragg's enthusiasm for gardening is well known, because it led him to engage in a subterfuge when he later moved to London. Having no plants to tend, he applied for and was employed as a part-time gardener, without revealing that the old bloke poking around in the shrubbery was a knighted Nobel Laureate.

One day while mixing liquids in a bucket in his Cambridge garden he noticed the floating raft of bubbles on its surface. It occurred to him that here was a splendid model for the atoms of a metal. This was during the distractions of World War II; the idea lay dormant for some years. Then he gave it to Nye, and they wrote a classic paper on the subject, familiar to a wide readership through Feynman's Lectures, where it is included.[2]

There is a curious parallel between this and Smith's idea. A metal grain is represented by a single flattened polygonal bubble in Smith's two dimensional foam. Bragg's bubbles are of equal size and small, so that they remain roughly spherical and may be taken to represent atoms, residing within grains. Within each grain they adopt the triangular structure that we encountered with the coins of Chapter 1.

---

[2] Bragg, W. L. and Nye, J. F., 1947, A dynamical model of a crystal structure, *Proc. Roy. Soc. (London)* A, **190**, 474.

# Chapter 11

# Toils and Troubles
with Bubbles

## 11.1 Playing with Bubbles

Smith and Bragg belong in a long line of bubble-blowers. Foams and bubbles have fascinated scientists of all ages, in all ages. Most have devoted some time to admiration of what Robert Boyle called "the soap bubbles that boys are wont to play with." Part of their charm and mystery lies in the colors produced by the interference of light in thin films. Small clusters of bubbles, or the extended ones we call foams (Figure 11.1), have elegant structures which call for explanation.

The painter, poet, and philosopher may point to the ephemeral nature of these things as a metaphor for our own mortality or the transience of fame and fortune, but the first interest of the scientist is in making relatively stable foams. This is not difficult with a little ordinary detergent solution shaken in a sealed container. Looking into it, one can see that, though disordered, it shows clear evidence of some principles of equilibrium at work. What are they?

When spherical bubbles pack together to form a foam (as when they rise out of a glass of beer) they are forced into polyhedral shapes as gravity extracts most of the liquid from their interstices. What began as a sphere packing, in conformity to the rules of earlier chapters, now presents a different paradigm for pattern in nature. In this case the density is fixed and surface area is to be minimized.[1]

---

[1] See for general reference: Sadoc, J. F. and Rivier, N. (eds.), 1999, *Foams and Emulsions* (Dordrecht: Kluwer); Ball, P., 1999, *The Self-Made Tapestry: Pattern Formation in Nature* (Oxford: Oxford University Press); Hildebrandt, S. and Tromba, A., 1996, *The Parsimonious Universe* (Berlin: Springer).

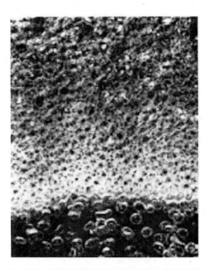

**Figure 11.1** Soap foam. (Courtesy of J. Cilliers, Imperial College.)

## 11.2 A Blind Man in the Kingdom of the Sighted

The man who most clearly saw what the principles of bubble-packing must be was blind. Joseph Antoine Ferdinand Plateau caused irreparable damage to his eyes by staring at the sun in an experiment on the retention of vision. The 1999 eclipse brought many public reminders of the extreme danger of doing so. He began to go blind in 1841 and had lost all vision by 1844.

Michael Faraday wrote consolingly and prophetically to him:

> Well may you and your friends rejoice that though, in the body, you have met with a heavy blow and great discouragement, still the spirit makes great compensation, and shines with glorious light across the bodily darkness.

Today, Plateau is remembered for his later researches, undertaken with the help of family, friends and students, leading to his great work *Statique Experimentale et Théorique des Liquides Soumis aux Seules Forces Moléculaires* (1873). An English translation of this is now available on Ken Brakke's Web site.[2]

As a hero of Belgian science, he was elevated to the rank of "Chevalier" in the Order of Léopold. His extraordinary dedication to science did not

---

[2] http://www.susqu.edu/brakke/.

**Figure 11.2** Playing with soap bubbles.

preclude a happy family life but he was often preoccupied. It is recorded that he disappeared for six hours while on honeymoon in Paris, returning eventually to his distraught bride, to say that he had forgotten that he had just been married. A similar case is to be found in the Irish mathematician George Gabriel Stokes.

At the heart of Plateau's classic text were those experiments with wire frames (Figure 11.3) dipped in soap solution which are still commonly used in lecture demonstrations. They were popularized as such by C. V. Boys in his *Soap-Bubbles, Their Colors and the Forces Which Mould Them, Being the Substance of Many Lectures Delivered to Juvenile and Popular Audiences*, published by the Society for Promoting Christian Knowledge, which seems

**Figure 11.3** A Plateau frame.

to have taken a broad view of the scope of such knowledge, in 1911.[3] The tradition of Plateau's entertaining experiments continues today in the hands of Cyril Isenberg[4] and others, and one can purchase the frames at modest cost from Beevers Molecular Models.

Plateau's book was well received. J. C. Maxwell reviewed it in *Nature*, first asking ironically—*Can the poetry of bubbles survive this?*—then replying with this encomium:

> Which, now, is the more poetical idea—the Etruscan boy blowing bubbles for himself, or the blind man of science teaching his friends to blow them, and making out by a tedious process of question and answer the condition of the forms and tints which he can never see?

The meaning of the book's title is not self-evident. It may be taken to mean *the laws of equilibrium of liquids under surface tension, when gravity is negligible.* Or, plainly put, *what are the shapes and connections of soap films?*

Plateau's laws which answer this question are as follows:

(1) Films can only meet three at a time and they do so symmetrically, so that the angles between them are 120°.

(2) The lines along which they meet are themselves joined in vertices at which only four lines (or six films) can meet. Again they are symmetric, so that the angle between the lines has the value $\cos^{-1}(-\frac{1}{3})$ or approximately 109° (the tetrahedral, or Maraldi, angle).

(3) The films and the lines are curved in general: the average amount by which the films are bowed in or out is determined by the difference in pressure between the gas on either side.

Note that the third law does not dictate that a zero pressure difference implies a flat film. Saddle-shaped surfaces can have zero mean curvature. The law, included here under the name of Plateau, really belongs to Laplace and Young, rivals of English and French science around 1800. They engaged in a bitter dispute for priority, exacerbated by the mutual disregard of the more practical English and the idealistic French. Young was engaged in another such Anglo–French competition, in the race to the translation of the Rosetta Stone. Today, we recognize the complementary contribution of Laplace and Young in the Laplace–Young law.

---

[3] A later edition is Boys, C. V., 1959, *Soap Bubbles* (New York: Dover).

[4] Isenberg, C., 1992, *The Science of Soap Films and Soap Bubbles* (New York: Dover).

If soap films are trapped between two glass plates in order to create a two-dimensional structure, only the first and third laws are needed.

The first two laws were inspired by observation and they could be rationalized by elementary theory up to a point. Soap films have energies proportional to their surface area, and they therefore tend to contract and pull with a force (the surface tension) on their boundaries. Plateau's laws express the conditions for stable equilibrium of these forces and the gas pressures which act on the films.

The first law is easily derived and extends our remarks about vertices in Chapter 7. The second is much more problematic.

## 11.3 Proving Plateau

Although he was not averse to mathematical analysis, Joseph Plateau's essential method was that of generalization from observation—he was content to leave it to others to find theoretical justifications of his laws.

Ernest Lamarle, a Belgian mathematician who was expert in differential geometry, provided some of this mathematical underpinning in the 1860s. He showed that the principle of minimal area implied everything that Plateau had reported. In particular, he gave a justification of the rule that states that only six soap films can meet at a point. A foam cannot have stable vertices formed by more than this number of films.

The proof runs to many pages. It begins with the classification of every possible form of vertex, consistent with the balance of the surface tension forces acting in the adjoining films. This in itself does not guarantee stability: most of these turn out to be unstable equilibria.

One might expect to encounter a great difficulty at this point, if there are infinitely many possibilities for such a vertex. But it turns out that there are only a few.

A small sphere centered on the vertex must have intersections with the soap films which form a pattern on its surface as follows (Figure 11.4):

(1) The lines are geodesics: that is, each lies in a plane which cuts the sphere in half.
(2) The lines intersect at 120°, three at a time.

The possible patterns of this type are highly reminiscent of the earliest stages in embryonic development.

Lamarle proceeded to devise ways in which each of the more complicated vertices could be deformed and dissociated into combinations of the elementary one, while lowering the total surface area. These define modes of instability of the vertex, hence disqualifying it as a possible stable configuration.

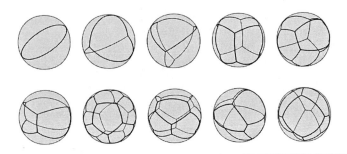

**Figure 11.4** Different equilibrium vertex configurations, in terms of geodesics. (Redrawn from Almgren, F. Jr. and Taylor, J., 1976, *Sci. Am.* **235**, 82–93. With permission.)

For serious mathematicians, the story does not end here. In particular, it reappeared on the agenda of Fred Almgren, 100 years after Lamarle.

Just as Lamarle had performed, at one level, a clean-up operation on Plateau's arguments, so Fred Almgren set out to perfect the proofs of Lamarle and others in the study of minimal surfaces.[5] Their work contained hidden or explicit assumptions of smoothness in the minimal structures that they purported to describe.

Mathematicians can conceive all sorts of strange entities with surfaces which are the opposite of smooth—not just rough, but perhaps infinitely so. In recent years, Benoit Mandelbrot has taught us that these monstrous constructions are not really alien to our world.

This said, it can hardly be maintained that much real doubt should be entertained about the assumption of smoothness in the particular case of soap films. Almgren's quest was for completeness and rigour, to eliminate nasty possibilities, however hypothetical, from the proof. Much of his fastidious work in this vein over a 35-year career at Princeton was compiled in a 1720-page paper, unpublished at the time of his death in 1997.

Another part of Almgren's legacy is to be found in successive generations of graduate students. One of these was Jean Taylor, who became his wife. It fell to her to construct the new version of Lamarle's proof, which she published in 1976.[6] It is of comparable length to the nineteenth century paper, but couched in the inscrutable language of geometric measure theory.

[5] Almgren, F. Jr., 1976, Existence and regularity almost everywhere of solutions to elliptic variational problems with constraints, *Mem. Am. Math. Soc.* **165**; Almgren, F. Jr. and Taylor, J., 1976, The geometry of soap films and soap bubbles, *Sci. Am.* **235**, 82–93.

[6] Taylor, J., 1976, The structure of singularities in soap bubbles-like and soap-film-like minimal surfaces, *Ann. Math.* **103**, 489–539.

Almgren and his successors have leavened their rather impenetrable studies with a lively sense of the more accessible and practical facets of their subject. In particular, Ken Brakke emerged from that school to write the Surface Evolver software, the fruits of which we will see later in this chapter.

---

### Minimal Surfaces

The theory of minimal surfaces has continued to be an active focus of research in this century. Several Fields Medals (a particularly prestigious mathematical prize) have acknowledged great achievements in that area. In particular, Jesse Douglas received the first medal in 1936 for his contribution to the "Plateau problem," which is concerned with a single soap film spanning a loop of wire of arbitrary shape.

---

## 11.4  Foam and Ether

Plateau's rules apply to any foam in equilibrium. They place restrictions on, but do not determine in full, the answer to our question: *Which structure is best?* This we have already answered for two dimensions; and, indeed, experiments with foams of equal bubbles do reproduce the honeycomb.

But just as ball bearings are uncooperative in the search for ideal structures, so are soap bubbles in three dimensions. This did not stop Sir William Thomson (Figure 11.5) (later Lord Kelvin) attacking the theoretical problem of the ideal ordered foam in 1887. He does not seem to have tried to make such a "monodisperse" foam, despite his credo (see below) that theory must be anchored in reality.

At that time Kelvin was the preeminent, if aging, figure of British science. He still had a strong appetite for scientific endeavour—in the end he published over 600 papers, a score worthy of the most competitive (and repetitive) of today's careerists. They stretch across the entire spectrum of physics from telegraphy and electrical technology to the second law of thermodynamics, for which we honor his name in the scientific unit of temperature. He also found time to advise the government, travel with the cable-laying ships and go sailing on his own yacht and, on one occasion, to make a proposal of marriage on the island of Madeira. Quite a man.

**Figure 11.5**  Sir William Thomson (Lord Kelvin) (1824–1907).

---

### Ether

Apollonius of Tyana is said to have asked the Brahmins of what they supposed the cosmos to be composed.

"Of the five elements."

"How can there be a fifth" demanded Apollonius "beside water and air and earth and fire?"

"There is the ether" replied the Brahmin "which we must regard as the element of which the gods are made; for just as all mortal creatures inhale the air, so do immortal and divine natures inhale the ether." [From Sir Oliver Lodge, 1925, *Ether and Reality* (London: Hodder and Stoughton), p. 35.]

---

One of the great quests of Victorian science was the search for a physical model for light waves. In the centuries-old debate between the advocates of particle and wave interpretations of light, the wave enthusiasts had gained the upper hand by the middle of the nineteenth century. It remained to specify the substance—the ether—the vibrations of which, like those of sound in air, constituted the light waves. We have already encountered the fanciful ideas of Osborne Reynolds, concerning the nature of the ether (Chapter 3). The word itself came down to us from the Greeks, for whom it represented a fiery heaven into which souls were received.

British natural philosophers were determinedly realistic in their outlook, as in the case of Young, always trying to relate the world of microscopic and invisible phenomena to everyday experience. This may be the reason that neither relativity nor quantum mechanics, both of which conflict with everyday experience, can be listed among the achievements of that school. Lord Kelvin, together with P. G. Tait, wrote in the preface to their textbook:

> Nothing can be more fatal to progress than too confident reliance on mathematical symbols; for the student is only too apt to take the easier course, and consider the *formula* and not the *fact* as the physical reality.

In common with others, Kelvin was not easily impressed by formalism and abstraction, despite being a first-rate mathematician. He did not even join the growing band of Maxwellians who fully accepted the theory of James Clerk Maxwell, in which light emerges not as a mechanical vibration but rather as a variation in electric and magnetic fields. The implication of the equations of Maxwell, which required a whole generation of debate to clarify, was triumphantly vindicated in 1887 by the experiment of Heinrich Hertz in Germany. In this climactic moment of the history of science, man "won the battle lost by the giants of old, has snatched the thunderbolt from Jove himself and enslaved the all-pervading ether." These are the words of George Francis Fitzgerald, addressing the British Association.

It was a little too late to convert Kelvin, who went on cooking up material ether models until he died, insisting that the elusive substance was "a real thing." "Nothing" said Fitzgerald later "will cure Sir William Thomson, short of the complete overthrow of the whole idea." Nothing indeed was going to deflect this gallant knight from tilting at a favorite windmill.

On September 29, 1887, Kelvin woke up, sat up, and wrote in his notebook *Rigidity of Foam*. He had conceived the notion that the ether might be a foam, a wild idea that Gibbs politely called "the audacity of genius," after complaining about the proliferation of published speculations on the ether. (Amazingly, the latest speculation on the nature of space–time now seems headed in the same direction.)

Kelvin turned to Plateau's book for inspiration in trying to decide what structure the foam should have, and was soon playing with wire frames. His niece Agnes King wrote on November 5th:

> When I arrived here yesterday Uncle William and Aunt Fanny met me at the door, Uncle William armed with a vessel of soap and glycerine prepared for blowing soap bubbles, and a tray with a number of mathematical figures made of wire. These he dips into the soap mixture and a film forms or adheres to the wires very beautifully and perfectly regularly. With some scientific end in view he is studying these films.

**Figure 11.6** Kelvin's palatial residence on the Scottish coast.

By then Kelvin had already solved the problem that he had set himself, that is, to define the ideal structure of equal bubbles: *What partitioning of space into equal volumes minimizes their surface area?* Or rather, he had come up with a reasonable conjecture, a masterful design which he thought nature must be compelled to follow. He recorded it in his notebook on November 4th.

Since Kelvin's first words on the subject—rigidity of foam—were written in early September at 7:15 A.M. while in bed, we may presume that his foam ether model was conceived during the night. This is a common enough phenomenon. The scientist who retires, his brain feverishly obsessed with a single problem, is likely enough to spend the night attacking it in that semiconscious state which is ideal for unbridled yet directed thought. Helmholtz gave as the requirement for mathematical reasoning that "the mind should remain concentrated on a single point, undisturbed by collateral ideas on the one hand, and by wishes and hopes on the other."

Maxwell described such an experience in a poem:

*What though Dreams be wandering fancies,*
*By some lawless force entwined,*
*Empty bubbles, floating upwards*
*Through the current of the mind,*
*There are powers and thoughts within us (. . .).*

Kelvin's thought indeed consisted of empty bubbles, for it was a foam without gas that he saw as a possible ether model. Such a thing cannot be stable but he convinced himself otherwise. So, when he had succeeded in describing an appropriate structure for the ether foam, he rushed into print with it in the *Philosophical Magazine*. The speed of this publication (not much more than a month later) rivals or exceeds that of the printed journals of today. It may have helped that the great man himself was editor of the journal, but one presumes that reviewers (if there were any) would hardly have questioned his insights.

## 11.5   The Kelvin Cell

The cell described by Kelvin may be described as a modified form of a truncated octahedron, a term used by Kepler. Kelvin chose to call it the "tetrakaidecahedron." Coxeter has called this name "outrageous"; and it does seem unnecessary. It was one of the 13 Archimedean solids and was familiar, for example, to Leonardo. It plays an important role in modern solid state theory and crystallography, but it was not so well known to physicists in the 1880s as it is today.[7] Kelvin's ability to visualize it derived in part from his contributions to early crystallography, to which he applied his characteristically down-to-earth approach, advising students to buy 1000 wooden balls and study their possible arrangements. He saw that this poly-hedron can be packed to fill all space. Furthermore, only a little curvature of the hexagonal faces is necessary to bring it into complete conformity with Plateau's requirements (Maraldi angles, etc.). He adroitly calculated this subtle curvature, "tweaked the model" as we would say today, but had difficulties in drawing it. True to form, he made a wire model so that it could be seen in tangible form. This survives in the University of Glasgow (which now has a Kelvin Museum) and is known as "Kelvin's Bedspring." (Figure 11.7), Visitors to what remains of the Lisbon EXPO site can admire the large Kelvin cells at the center of a framework of cables, designed as a climbing frame for children.

Kelvin's various attempts to bring forth an acceptable model for the material ether were greeted with limited and diminishing enthusiasm. This particular version was called "utterly frothy" by a Cambridge don. When Fitzgerald tried to be conciliatory by suggesting that Kelvin's models were at best allegories, he received a spirited retort from the great man, in the words of Sheridan's Mrs Malaprop: "Certainly not an allegory on the banks of the Nile."

It is curious that Kelvin said nothing about a direct experimental test, although he described a way of making his structure with wire frames

---

[7] See: Weaire, D. (ed.), 1997, *The Kelvin Problem* (London: Taylor & Francis).

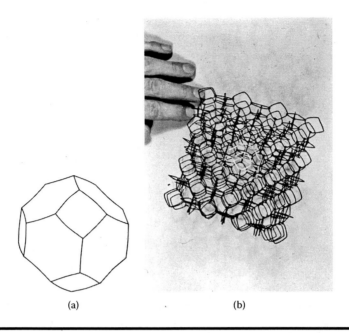

**Figure 11.7** (a) The Kelvin cell. (b) "Kelvin's Bedspring." (Courtesy of the University of Glasgow.)

(which begs the question). Perhaps he was unaware of the ease with which equal bubbles can be made, simply by blowing air through a thin nozzle immersed in a soap solution. Unfortunately, his unconfirmed conjecture was accepted too readily as the established truth by others. It was as if, as John Ziman once said of another theoretical model, "the Word had been made Flesh." Uncritically accepted, his conjecture remained unchallenged for quite some time.

In one unfortunate sequel, a Russian mathematician posed the same problem as Kelvin, and developed the same conjecture, in 1992. He was apparently oblivious of the work of his illustrious predecessor.

## 11.6  Most Beautiful and Regular

There has never been much doubt that Kelvin's is the correct solution if all foam cells are restricted to have an identical shape and orientation. The doubt arises when they are given more freedom than this, as they have in nature, while maintaining equal volumes.

It is part of the physicist's faith that things are simple. (But not too simple, as Einstein warned.) There is always a provisional prejudice in favor of a neat solution rather than a complicated one. Or, put in grander language by D'Arcy Wentworth Thompson, "the perfection of mathematical beauty is such (as Colin MacLaurin learned of the bee), that whatsoever is most beautiful and regular is also found to be most useful and excellent."

Such a precept would be greatly improved by the addition of "generally speaking" but the imperious sweep of this superlative prose stylist would not admit it. The Kelvin problem was for him solved "in the twinkling of an eye," presumably that of Kelvin. Thus was the Word made Flesh, at least in the mind of the succeeding generation.

Kelvin's favorite polyhedron also played a large role in his research on space-filling structures in relation to general crystallography. In 1893, he wrote to Rayleigh about this, in the midst of filing patents, worrying about Home Rule, and other preoccupations. Referring decorously to his wife as "Lady Kelvin," as befits a correspondence between two members of the House of Lords, he said that she had begun to make a tetrakaidecahedral pin-cushion, which "will make all clear." Kelvin always brought his work home.

## 11.7 The Twinkling of an Eye

The American botanist Edwin Matzke used Thompson's phrase as the title of a lecture he gave to the botanical club at Columbia University in 1950. He poured scorn on the widespread acceptance of Kelvin's cell, together with the erroneous conclusions of Buffon and Hales, also repeated uncritically by Thompson. They were relegated to the "limbo of quaint and forgotten dreams."

It had led him and other biologists to undertake fruitless searches for the Kelvin cell in natural cellular structures. It was nowhere to be found. He had been driven to perform the experiment so long overlooked, by making an actual foam of equal bubbles.

He did so in a manner which now looks foolish. It should have been known to him that such bubbles may be created simply by blowing air steadily through a fine nozzle beneath the surface of a soap solution. Instead he and his assistants blew every bubble individually with a syringe and added it carefully to the foam.

The result was a disordered structure (like the bag of ball bearings in Chapter 5). Matzke examined his mass of bubbles meticulously with a binocular microscope, in the manner of a biologist, preparing drawings of many cells for publication. His careful scrutiny revealed not a single Kelvin cell.

Very recently, simulations by Andrew Kraynik have reproduced Matzke's results for the statistics of occurence of different kinds of cells. Any doubts about Matzke's experimental procedure (did he really avoid coarsening) were allayed.

"Is this an indictment of twinkling eyes?" asked Matzke in the end, referring to what Thompson had said in his unconditional acceptance of Kelvin cell, and generously answered, "no"—but not before condemning the naivety of his predecessors.

Matzke's heavy labor and even heavier irony discouraged others from any further seach, but the mathematical question remained. Was Kelvin's solution the best, at least in principle? Some authorities, for example Hermann Weyl, were inclined to believe that it was, while others had an open mind. Fred Almgren of Princeton said in 1982 that "despite the claims of various authors to the contrary it seems an open question." A few mathematicians[8] and especially the Almgren group kept the Kelvin problem alive. One of Almgren's graduate students (John Sullivan) has said: "I think most people assumed this partition was best. But Almgren alone was convinced it could be beaten."

Rob Kusner maintains that many other people were also convinced that it could be beaten. Indeed he was among them. In 1992, he proved that an assembly of equal-pressure bubbles can minimize the interfacial surface area in a foam with an average number of faces greater or equal to $13.39733\ldots$[9] The inequality did not exclude the Kelvin solution, but Kusner (following ideas from Coxeter and Bernal) was inclined to think that a minimal partition must have an average number of faces close to 13.5.

The problem remained: The Kelvin Problem.[10]

## 11.8 Simulated Soap

Today's mathematicians can ease their frustration with the solution of difficult problems and with the illustration of abstract results by the use of powerful computer simulations and advanced graphics. A leading exponent of such techniques is Ken Brakke of Susquehanna University in Pennsylvania. A former student of the Princeton school of Fred Almgren, he set out to develop a new and flexible computer code for producing surfaces of minimum area (Brakke's Surface Evolver). Once completed, it was generously offered to the world at large.[11] Figure 11.8 is an example of its application

---

[8] Choe, J., 1989, On the existence and regularity of fundamental domains with least boundary area, *J. Diff. Geom.* **29**, 623–663.

[9] Kusner, R., 1992, The number of faces in a minimal foam, *Proc. R. Soc.* A **439**, 683–686.

[10] Weaire, D. (ed.), 1997, *The Kelvin Problem* (London: Taylor & Francis); Rivier, N., 1994, Kelvin's conjecture on minimal froths and the counter-example of Weaire and Phelan, *Phil. Mag. Lett.* **69**, 297–303.

[11] http://www.susqu.edu/FacStaff/b/brakke/.

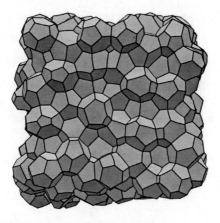

**Figure 11.8**  Simulation of foam bubbles obtained with Evolver starting from the Voronoï partition in Figure 7.4. (Courtesy of Andrew Kraynik.)

to foams. It has been continuously updated ever since, and used for many things, from the shape of a pendant liquid drop to the modeling of solder connections in semiconductor circuits. It is, in particular, ideal for the Kelvin problem and it was set to work on it around 1990.

Kelvin's conjecture at first survived this first onslaught by modern technology: no better structure could be found.

## 11.9   A Discovery in Dublin

In late 1993, Robert Phelan began his research at Trinity College Dublin (Figure 11.9). His task was to explore the Kelvin problem and variations upon that theme, using Brakke's Surface Evolver.

Phelan had joined a computational physics group which had a broad background in solid state and materials science. Hence there was no question of a blind search for an alternative structure. An idea was formed of

**Figure 11.9**   Denis Weaire and Robert Phelan in 1993.

the type of structure which might be competitive with that of Kelvin, essentially one with a lot of pentagonal faces. What structures in nature have such a form?

There is one class of chemical compounds in which covalent bonds create suitable structures: the bonds are tetrahedral. The compounds are called clathrates, a reference to the fact that they are made up of polyhedral *cages* of bonds. Usually, they form because they create convenient homes for guest atoms or molecules. Gas pipelines in the Arctic are sometimes clogged up with clathrate crystals of ice.

The cages of bonds can be visualized as foam cells. Most of the rings of bonds on the sides of the cages are fivefold, creating pentagonal faces, as seemed to be required. The suggestion therefore was to explore the clathrates, particularly the two simplest ones, Clathrate I and II. The first of these was fed into the Evolver as a real foam structure with equal cell volumes. Such a structure is a regular assembly of two types of cell with, respectively, 12 and 14 faces and results in a foam with 13.5 faces in average (Figures 11.10 and 11.11).

When the first output emerged, it was immediately evident that it was going to defeat Kelvin. When fully equilibrated, it turned out to have a surface area 0.3% less than that of the venerable conjecture. This does not sound like very much but endeavours in optimization always have to strive

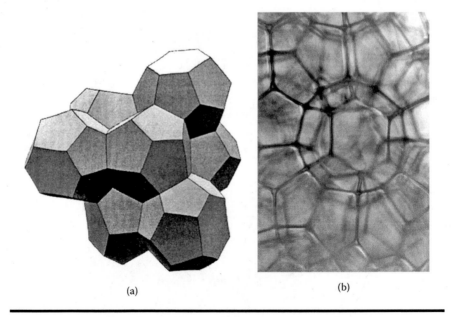

(a)  (b)

**Figure 11.10** (a) The Weaire–Phelan structure. (b) The observation of these cells in a real foams.

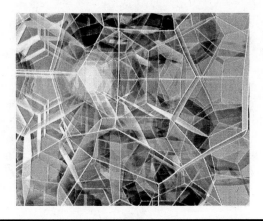

**Figure 11.11** The Weaire–Phelan structure in a computer-generated image by J. M. Sullivan. (With permission.)

for quite small differences, in horse-races and elsewhere. The margin of success in this case was recognized as quite large.[12]

```
Date: 9 December 1993 03:25:03.13
From: "brakke@geom.umn.edu"
Reply-To: "brakke@geom.umn.edu"
To: "dweaire@vax1.tcd.ie", "rphelan@alice.phy.tcd.ie"
cc:
Subject: Kelvin

I confirm your results. I got the area down to 21.156.
As soon as I saw the picture on the screen, I was
sure you had it. I always figured the way to beat
Kelvin was to use lots of pentagons, since pentagon
vertex angles are very close to the tetrahedral angle.
So when I saw you had almost all pentagons, I knew
you'd done it.

Congratulations.

Ken Brakke
```

It was, according to Almgren, a "glorious day for surface minimization theory" but, strictly speaking, it provides no more than a counterexample

---

[12] Later, Almgren, Kusner, and Sullivan provided a rigorous proof that the Weaire–Phelan structure is of lower energy than that of Kelvin.

to Kelvin's conjecture and hence a replacement for it. The problem of proof that it is optimal in an absolute sense remains, though confidence grows that this structure will not in turn be surpassed: many related candidates have already been tried, mostly from the wide class of clathrate structures.

The Kelvin Problem has thus acquired the status long held by the Kepler Problem: something we "know" but cannot prove. If you look forward to this proof, don't hold your breath: it would not be found in this (or any other) edition of the present work. This is a yet more demanding task for the computer and its programmer than those provoked by Kepler. Some day a largely autonomous computer may give the answer. One wonders what the final output will be.

Perhaps just YES, after a year's silence?

## 11.10   Crystals of Small Bubbles

Bragg's two dimensional soap raft (Section 10.3) has remained well known. Nevertheless, a discovery that he and his student reported at the end of their paper somehow escaped attention until now. They found that their small bubbles crystallized when they formed a three-dimensional mass (Figure 11.12).

Dry foams, as considered by Kelvin, are reluctant to form crystals, even when the bubbles are of equal size. The story is different when the foam is "wet," consisting of almost spherical bubbles, like the hard spheres of earlier chapters. Bubbles which are small enough will remain spherical under the force of gravity, which is not strong enough to distort them as they pack together.

The behavior of these small (0.1 mm diameter) bubbles is paradoxical. They are not really small enough to be subject to thermal motion: they cannot search for their best packing, to reduce their energy under gravity. Colloid particles can do this, but they are much smaller spheres and they need to settle down very slowly. On the contrary, these small-bubble crystals form extremely quickly.

The crystal structures that are found are all close-packed: cubic close packed (fcc) and the other variants listed in Chapter 13. There seems to be a definite preference of fcc. Viewed from above the crystals mostly show the same triangular pattern as the two dimensional raft. But in addition to these close-packed planes, (100) fcc surfaces are seen as square patterns, as in Figure 11.12a. A careful look through a microscope reveals at least the first three layers, but in a higly distorted form.[13]

---

[13] Van der Net, A., Delaney, G. W., Drenckhan, W., Weaire, D., and Hutzler, S., 2007, *Colloids and Surface A-Physicochemical and Engineering Aspects*, in print.

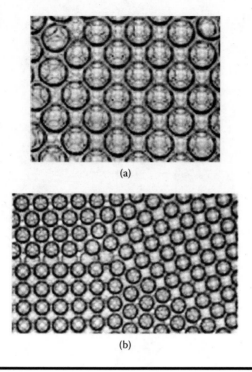

(a)

(b)

**Figure 11.12**   Bubble crystals.

Apart from the fascinating detail of dislocations, grain boundaries and defects, this spontaneous order should have attracted continued interest. It seems at odds with the behaviour of dry foams on the one hand and hard spheres on the other. One might well have expected the see the Bernal dense random packing because these bubbles are small enough (0.1mm) to remain spherical when pressed together by gravity.

There can be no thermal motion of particles on this scale: How then do they form a crystal?

This remains a mystery, together with Bragg's failure to follow such a promising avenue of research. His work on the two-dimensional raft had been held up by the war for several years. Having finally completed it, perhaps he felt that it was time to move on.

*Chapter 12*

# Bubbles in Beijing

## 12.1 An Olympian Vision

In preparation for the Beijing Olympic Games of 2008, the Chinese government announced a competition for the design of its principal buildings. One was to be the Aquatic Center.

The winning proposal, submitted by PTW Architects and the Arup Australasia engineering group, together with CASEC (China), was an extraordinary feat of the imagination, but not of a purely arbitrary kind. It sought to draw from science a hint of something that might be iconic and emblematic in relation to water itself. Why not bubbles? After all, the guiding spirit of the Arup Corporation, Sir Ove Arup, "in seeking to achieve the perfect union of design and construction," has been described as "continually throwing out colored bubbles."

## 12.2 Fun and Fit for Purpose?

Bubbles in architecture do have antecedents, particularly in the work of Frei Otto. But here they take a startling new form—the Water Cube. The building consists of thick slices of the Weaire–Phelan foam of the last chapter. It is adjusted to tidy up its termination on the walls and roof, and canted at such an angle as to give the impression of a more disordered structure, reminiscent of the foams of everyday life, but leaving engineers with the comfort of a regularly repeating construction. The designers had begun with the idea of using Kelvin foam structure (see Chapter 11), but it appeared much too regular to the eye.

The edges (Plateau borders) of the foam are represented by huge steel beams, 90 kilometers in all, heavier at the bottom, lighter at the top. To hoist

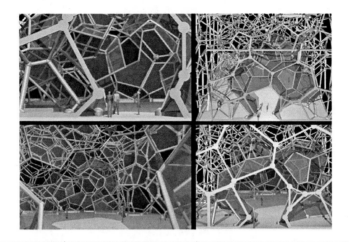

**Figure 12.1**    The Water Cube, restaurant area. (Artist's impression. Courtesy of Arup Group.)

these into position and join them must have been a formidable challenge, but once mastered it followed a simple recipe. The more random main stadium that stands beside it (nicknamed the "Bird's Nest") seems to have posed more problems for its builders.

The outer and inner skins of the building display two-dimensional cellular patterns which are cross-sections of the structure. They convey an impression of a foam surface, especially since their cells are covered by convex plastic cushions of the same kind used in the pioneering Eden project in southwest England. So the walls present two layers of air, which can be managed to control heat and temperature, while contributing solar heating, and helping the design to win a major prize at the Venice Biennale.

The restaurant space (Figure 12.1) provides a particularly striking impression. It is created by removing cells from the structure, leaving an irregular faceted surface.

## 12.3   A Flexible Friend?

Some engineers may at first have cocked a sceptical eyebrow at this design. From an early stage their education has always inculcated the virtues of triangulation, the creation of rigid units of struts that make for stability and strength. The interior of the Water Cube walls contains no triangles, such as we would find if we joined the points of (for example) the face-centered cubic lattice. Instead, straight steel beams (whose counterparts would be slightly curved in a real foam) meet four at a time in an open structure that looks and is relatively flexible.

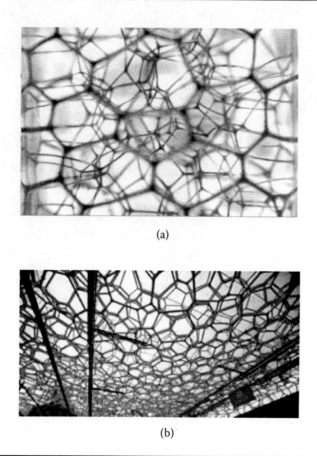

(a)

(b)

**Figure 12.2** (a) A liquid foam in a glass container: this is the general impression conveyed by the Water Cube design (b).

No wonder then that a massive effort went into validation of the unconventional design. Today's computer power frees the architect from many of the cautious constraints of the past.

Simulations showed the design to be safe and satisfactory, and indeed to have one precious virtue, in that its very flexibility made it inherently resistant to damage from seismic disturbance, a significant risk in China.

The theory of the stiffness of frames goes back to James Clerk Maxwell in 1864. He is more well known for his superlative contributions to electromagnetic theory and thermodynamics, but we should hardly be surprised by his contributions to the mathematics of mechanics, a prime ingredient of a Cambridge education.

Maxwell made a distinction between two classes of frameworks of narrow rigid struts, freely jointed, according as they did or did not allow relative movement of their parts. In the jargon of engineering the first type is a "structure" (typically a triangulated frame) while the second is a "mechanism."

Although a real structural framework is made of struts that are not completely rigid, nor are they usually freely jointed, Maxwell's two classes still have relevance. A "mechanism," rendered stable by the introduction of rigid joints, becomes rigid but it supports a load by the bending of its struts, instead of by tension/compression. Such a framework, of which the Water Cube is an example, has a relatively large elastic displacement under a load, because much less force is needed to bend a beam rather than compress it.

There is a type of biological structure that looks remarkably like the Water Cube, in that it consists of tetrahedrally-branched repeating units. The "prolamellar body membrane lattices" are ordered in a variety of ways,[1] including various clathrate structures such as that which was adapted for the Weaire–Phelan structure and the Water Cube. What biological principle selects these structures under different conditions remains unknown.

---

[1] Gunning, B. S. and Steer, M. W., 1996, *Plant Cell Biology* (Jones and Bartlett: Boston), p. 26.

# Chapter 13

# The Architecture of the World of Atoms

## 13.1  Molecular Tactics

Another book which appeared in the same year as that of Plateau (1873) was *L'architecture du Monde des Atomes* by Marc-Antoine Gaudin, from which we borrow the title for this chapter on the role of packing ideas in crystallography.

Gaudin sought to reconcile the laws of chemistry with the findings of early crystallography, the experimental part of which comprised the study of the external forms of crystals. He constructed molecules of various shapes, consistent with the symmetry of the corresponding crystal. The molecules were composed of atoms in the required proportions. These atoms were, in turn, considered to be made up of particles of ether but it sufficed for his purpose that they were assumed to be roughly spherical and packed together with roughly constant separations. His illustrations were delightful (Figure 13.1) but his speculations were no more than a shot in the dark, one small chapter in a confused story not concluded until the early twentieth century.

That crystals owe their beautiful angular forms to regular arrangements of atoms or molecules was a very old hypothesis, but only in the late nineteenth century was it at last pursued with rigor and related to properties: This was the birth of solid state physics, which grew to be the dominant sector of modern physics, at least in terms of the active population of researchers,[1] and industrial applications.

---

[1] See for general reference: Smith, C. S., 1981, *A Search for Structure* (Cambridge, MA: MIT Press); Burke, J. G., 1966, *Origin of the Science of Crystals* (Berkley, CA: University of California Press); Weaire, D. L. and Windsor, C. G. (eds.), 1987, *Solid State Science, Past, Present, and Predicted* (Bristol: Institute of Physics).

**Figure 13.1** Gaudin's drawings of hypothetical molecular structures (1873).

Identifying what Kelvin called the "molecular tactics of a crystal" remained a hesitant and erroneous process until x-rays provided the means to determine these arrangements reliably if not quite directly, early in the twentieth, century. Today, we can finally see (or, more accurately, feel) the individual atoms on the surface of a crystal, using the scanning tunneling or atomic force microscope.

## 13.2 Atoms and Molecules: Begging the Question

Whether matter is discrete or continuous has been a subject of debate at least since the time of the ancient Greek philosophers. The first detailed atomistic theory was that of Plato (in the *Timaeus*) who described matter to be "One single Whole, with all its parts perfect." He associated the four "elements"—*earth, water, fire* and *air*—with the form of four regular polyhedra—*cube, icosahedron, tetrahedron* and *octahedron*—and with the *dodecahedron* he associated the *Universe* (Figure 13.2). Plato attempted to match the properties of the elements with the shapes of the constitutive atoms. For instance water, being fluid, was associated with the icosahedron, which is the most spherical among the five regular solids. This theory was able to offer an *ad hoc* explanation of phase transitions (for instance the transition solid-to-liquid-to-vapor, which means earth-to-water-to-air and corresponds to cube-to-icosahedron-to-octahedron).

The atomistic theory of Plato was dismissed by Aristotle. He argued that if the elements are made up of these particles then the copies of each regular polyhedra must fill the space and this operation cannot be done with the icosahedron and the octahedron (he thought erroneously that it was possible to fill space with regular tetrahedra).

In other periods atoms have been reduced to mere points, or considered to be hard or soft spheres, or to have more exotic shapes endowed with specific properties inspired by chemistry. As Newton said, the invention of such "hooked atoms" by followers of Descartes often begged the question.

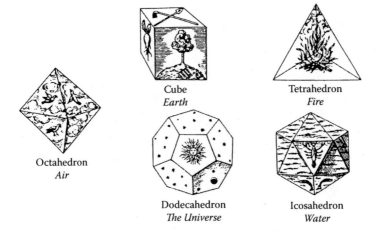

Cube
*Earth*

Tetrahedron
*Fire*

Octahedron
*Air*

Dodecahedron
*The Universe*

Icosahedron
*Water*

**Figure 13.2** The four elements and the Universe in Plato's conception. (From a drawing by J. Kepler. With permission.)

Christian Huygens in his *Traité de la Lumière* (1690) suggested that the Iceland spar (at that time very much studied for its intriguing optical property of birefringence) may be composed of an array of slightly flattened spheroids (see Figure 13.3c). In this way he explained at one stroke both the birefringence and the cleavage properties of these crystals.

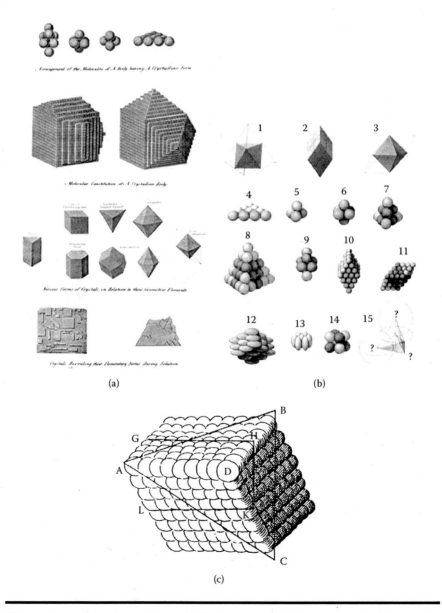

Figure 13.3 Various atomic models for crystals: Haüy (a), Wollaston (b), Huygens (c).

Haüy's celebrated constructions (Figure 13.3) begged the question, in as much as he explained the external form of crystals as being due to the packing of small components which were identical to the crystal itself. But his 1784 *Essai D'une Theorie sur la Structure des Crystaux* was one of the most perspicacious of early attempts to make sense of crystals. He recognized that their angles are not arbitrary but follow certain rules, still used today.

## 13.3 Atoms as Points

The Newtonian vision was taken to an extreme by Roger Joseph Boscovitch, early in the eighteenth century. He postulated that "matter is composed of perfectly indivisible, nonextended, discrete points," which interacted with one another.

Boscovitch published his *Theory of Natural Philosophy* on 1758, when he was a professor at the Collegium Romanum. He has been described as a philosopher, astronomer, historian, engineer, architect, diplomat, and man of the world. Given all this, the book is a disappointingly dry exposition in which he attempts to deduce much of physics from a "single law of forces." This means a mutual force between each pair of points. Boscovitch struggled to describe a possible form for this, drawing illustrations which resemble modern graphs of interatomic interactions.

Half a century later, French mathematicians such as Navier and Cauchy developed powerful theories of crystal elasticity based on this idea but independent of any particular form for the interaction. Their elegant analysis of the effect of crystal symmetry on properties has provided one of the enduring strands of physical mathematics. However, they were little concerned with the origins of crystal structure itself.

Late in the nineteenth century, the Irish physicist Joseph Larmor reduced the role of matter still further, to a mere mathematical singularity in the ether. His view was enshrined in the book *Aether and Matter*, to which contemporaries jokingly referred as "Aether and No Matter." It was published in 1900, a date after which Larmor declared that all progress in physical science had ceased. He was probably not serious—it is, in fact, the date of the inception of quantum theory, which finally told us what atoms are really like.

Today's quantum mechanical picture of a nucleus surrounded by a cloud of electrons is a subtle one: it taxes the resources of the largest computers to predict what happens when these clouds come into contact. The day has not yet come when older, rough-and-ready descriptions are completely obsolete. It is still useful for some purposes to picture atoms as hard balls with relatively weak forces of attraction pulling them together. In particular, the structures of many metals can be understood in this way.

**Figure 13.4** All snow crystals have a common hexagonal pattern and most of them show a hexagonal shape.

## 13.4 Playing Hardball

Among the manifold older ideas about atoms, the elementary notion of a hard sphere has endured as a useful one, even today.

Spherical "atoms" were adopted by Kepler to explain the hexagonal shape of snowflakes (Chapter 3). He did not consider these spheres to be atoms in the modern sense but more as the smallest particle of frozen water.

Of the later reinventions of this type of theory, one of the most influential was that of William Barlow, writing in *Nature* in 1883.

Barlow is a prime example of the self-made scientist. His paper "On the Probable Nature of the Internal Symmetry of Crystals" is remarkable for its total lack of any reference to previous work. He happily ignored centuries of speculations, particularly those of his compatriot William Wollaston (see Figure 13.3) early in the same century. His influence, particularly on and through Lord Kelvin, may be attributed in part to his clear writing style, his choice of the simplest cases and his use of attractive illustrations (Figure 13.5).

Barlow's method was to look for dense packings of spheres, with no attempt at proof that they were the densest. This he left for mathematicians to consider later. He mentioned the possibility that soft spheres might make more realistic atoms but did little with this.

His down-to-earth common-sense approach might be regarded as a reaction against the refined mathematics of the French school, which many British natural philosophers, such as Tait, had found rather distasteful.

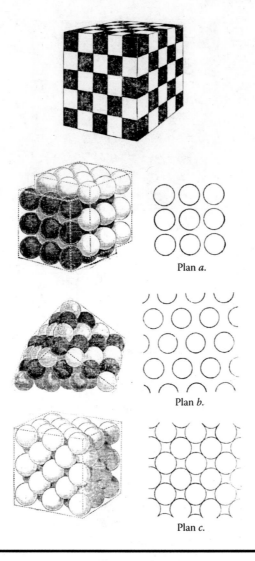

Plan *a.*

Plan *b.*

Plan *c.*

**Figure 13.5**    Some illustrations from Barlow's papers.

John Ruskin, venturing from the arts into science in a manner which was then fashionable, had in his *Ethics of the Dust* (1865), which consisted of "ten lectures to little house wives on the elements of crystallization," declared that the "mathematical part of crystallography is quite beyond girls' strength." One might suppose that it would be beyond Barlow's strength as well, but eventually he took it up avidly, and absorbed the full mathematical theory of the French, in later years. Indeed, he published in that area (albeit with some further disregard for the precedents).

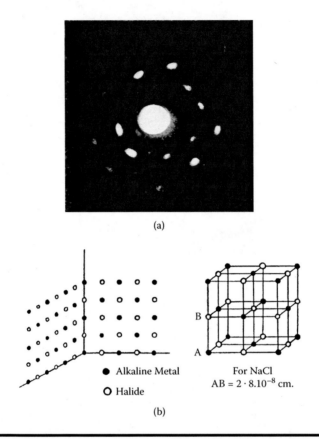

(a)

(b)

**Figure 13.6** X-ray diffraction pattern (a) and the relevant crystalline structure (b) for *NaCl*, as reported by Bragg (1913). (With permission.)

By 1897, he was ready to expound a more mature version of the theory in a more erudite style. He described many possible stackings for his hard spheres, of equal or unequal sizes.

Barlow's intuitive attack scored a number of notable hits, particularly in predicting the structures of the alkali halides, such as *NaCl*. (See Figure 13.6.) His place in the history of science was then assured by a contemporary and apparently unrelated discovery.

On November 8, 1895, a professor of physics in Würzburg, W. C. Röntgen, realized that a new type of ray was emanating from his discharge tube. It is said that he called his mysterious rays "x," in the hope that in the first citation they would be called by his name, as was usual in those times. The "Röntgen ray" was an immediate popular sensation, but the press used the name proposed by the discoverer and they passed into history as x-rays, much to Röntgen's distaste. The potential for medical science (Figure 13.7) and the challenge to the modesty of Victorian ladies were

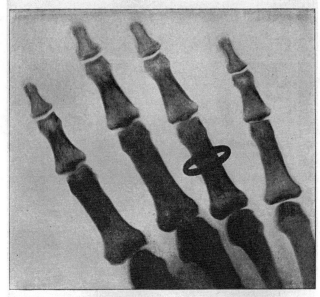

RADIOGRAPHIE DES OS DE LA MAIN
OBTENUE AVEC UN TUBE A FOYER DE PLATINE.
Épreuve de M. J. Chappuis.

**Figure 13.7**    One of the earliest x-ray photographs (1896).

clear—the implications for physics were not. One of the most extraordinary of these, which took two decades to emerge, was the determination of crystal structures using x-rays (see Section 13.5), vindicating much of the guesswork of Barlow and others. This was provided by William Henry Bragg and William Lawrence Bragg, father and son.

It should not be thought that all crystal structures are dense packings of balls. In the structure of diamond (Figure 13.8), each atom has only four atoms as neighbors. And this does not even qualify as a stable loose packing. This structure too was presaged before it was observed—this time by Walter Nernst (1864–1941), better known for his Heat Theorem (the Third Law of Thermodynamics).

In the late nineteenth century, van't Hoff's explorations of "Chemistry in Space"[2] gave us our modern conception of the molecule as a geometrical object, in which atoms take characteristic preferred arrangements, dictated by their chemical bonds. In particular, a set of four bonds was a possibility for carbon.

His crucial insight was not immediately popular with everyone. Kolbe's reaction was as follows: "(...) a miserable speculative philosophy, useless

---

[2] van't Hoff, J. H., 1875, *Chimie dans l'Espace.*

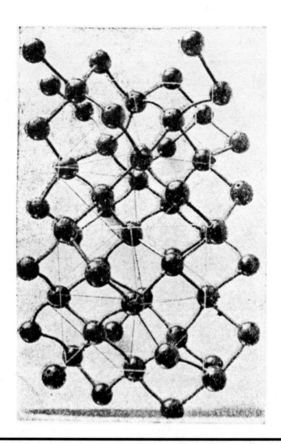

**Figure 13.8** Walter Nernst's 1912 model for diamond.

in reality, while apparently deep and ingenious, is springing up like a rank growth.(...) It is again being rescued from the lumber-room of man's erratic speculations, by certain pseudo-philosophers seeking to thrust it forward surreptitiously, like some fashionably and gorgeously attired female, intruding into good society which is not her place."

The unwelcome intrusion has stayed with us ever since. However important may be the more elusive description of quantum mechanics, chemistry and much of physics remains concerned with how simple basic units can be fitted together.

## 13.5 Modern Crystallography

In a crystal the structure repeats a local configuration of atoms as in a three-dimensional wall paper. There are only 14 ways to construct such periodic structures in three dimensions, the *Bravais lattices*, but they differentiate into 230 different types of internal symmetry. This "crushingly high number of 230 possible orderings," as Voigt called it, was both challenging and depressing to the theorist, until x-ray diffraction offered the means to use the theory in every detail.

When x-rays (electromagnetic radiation with a typical wavelength between 0.1 and 10 Ångstroms, i.e., between 0.000 000 01 and 0.000 001 mm) are incident on a crystal, they are diffracted and form a pattern with sharp spots of high intensity corresponding to specific angular directions. On June 8, 1912, at the Bavarian Science Academy of Munich, a study entitled "Interference effects with Röntgen rays" was presented.[3] In this work, Max von Laue developed a theory for the diffraction of x-rays from a periodic packing of atoms, associating the spots of intensity in the diffraction pattern with the regularity of the positions of the atoms in the crystal structure. Just one year later the elder Bragg reported the first determination of crystal structures from x-ray diffraction for such compounds as *KCl*, *NaCl*, *KBr* and *KI* (Figure 13.8), confirming Barlow's models.[4]

## 13.6 Crystalline Packings

In many crystal structures atoms are in positions corresponding to the centers of spheres in a sphere packing, as Barlow had supposed. It has been noted by O'Keeffe and Hyde,[5] two experts in crystal structures, that "it is hard to invent a simple symmetric sphere packing that does not occur in nature."

Of these, the most important family is that of packings with the maximal density $\rho = 0.740\,48\ldots$ of which Kepler's cubic close packing (or face-centered cubic, fcc) is a member. This maximal density can be realized in infinitely many ways, all of which are based on the stacking of close-packed layers of spheres (Figure 13.9), as practiced by the greengrocer.

But *two* possibilities present themselves for the relative location of the next layer, if it is to fit snugly into the first one. Each successive layer offers a similar choice, and only by following a particular rule will the fcc structure emerge. A different rule produces the hexagonal close-packed (hcp) structure, also described by Barlow, the next in order of simplicity. These two structures occur widely among the structures formed by the elements of the Periodic Table. More complex members of the same family—for example the double hexagonal close-packed structure—are found in alloys.

[3] Friedrich, W., Knipping, P., and Laue, M., 1912, Interferenz-Erscheinungen bei Roentgenstrahlen, *S. B. Bayer. Akad. Wiss.*, pp. 303–322.

[4] Bragg, W. L., 1913, The structure of some crystals as indicated by their diffraction, *Proc. Roy. Soc.* A **89**, 248–277.

[5] O'Keeffe, M. and Hyde, G. B., 1996, *Crystal Structures* (Washington, DC: Mineralogical Society of America).

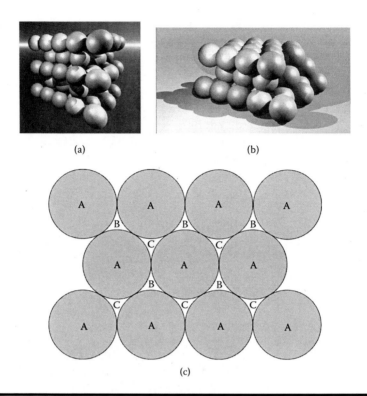

(a)                                    (b)

(c)

**Figure 13.9**   The hexagonal closed-packed [hcp] (a) and cubic closed-packed [fcc] (b) structures. These structures are generated by a sequence of layers of spheres in the triangular packing configur ation (c). Suppose that the first layer of spheres has centers in position A, the second layer can be placed in position B (or equivalently C), and for the third layer we have two alternatives: (i) placing the centers of the spheres in position A generating the sequence ABABAB... (which corresponds to the hcp structure); (ii) placing the centers in position C generating the sequence ABCABCABC... (which corresponds to the fcc structure).

If the packing fraction is decreased a little from its maximum value, allowing the hard spheres some room to move, and they are given some kinetic energy, what can be said about the competition between these structures? This is a very delicate question for thermodynamics, and it has only been settled recently by extensive computations. The winner (as the more stable structure) is fcc.[6]

The original appeal of crystals lay in their external shapes, and these provided clues to their internal order. However, the precise shape of a crystal in equilibrium cannot be deduced from this order alone. According to

---

[6] Woodcock, L. V., 1997, Entropy difference between the face-centered cubic and hexagonal close-packed crystal structures, *Nature* **385**, 141–143.

a principle enunciated by Gibbs and Curie in 1875, the external shape of a crystal minimizes the total surface energy. This is made up of contributions from each facet but different types of facet are more or less expensive in terms of energy.

Some of the observed shapes (for example, a perfect cube) can be realized by a rule developed by Bravais: *The largest facets have the densest packing of atoms* (which might be expected to have the lowest surface energy).

We may choose not to let thermodynamics dictate the natural shape, but cut the crystal to suit our taste, as in the diamond trade. But the facets must still correspond to planes of atoms, with special angles between them, accordingly to the old rule of Haüy. Much of a diamond's symmetry is therefore due to its internal order. The perfection of this symmetry can be beautifully demonstrated in a darkened room, with a small laser and a ring borrowed from the audience in the best conjuring tradition. When the laser is pointed to the crystal, a beautiful pattern emerges on the ceiling and walls. The laws of optics allow the light to escape only at right angles to the facets, in beams that reveal their perfect symmetry.

Some natural elements which have an fcc or hcp crystalline structure are given in Table 13.1.

## 13.7   Packing Tetrahedra

We have repeatedly remarked that regular tetrahedra cannot pack together to fill space but irregular ones may do so. Of special interest for crystal chemistry are packings in which neighboring atoms are on the vertices of such a system of closely packed tetrahedra. These structures are called "tetrahedrally packed."

A very important tetrahedrally-packed structure is the *body-centered cubic lattice* (bcc) (Figure 13.10). This is the crystalline structure of many chemical elements. The bcc structure is the only tetrahedrally packed structure where all tetrahedra are identical.

A special class of structures consists of those in which the packing is restricted to configurations in which five or six tetrahedra meet at each edge. These are the crystal structures of some of the more important intermetallic phases. Such structures were described in the 1950s by Frank and Kasper.[7] These two eminent crystallographers were inspired to produce their classification of complex alloy structures by a visit to Toledo, where the Moorish tiling patterns incorporate subtle mixtures of coordination and symmetry, particularly fivefold features such as the regular pentagon.

---

[7] Frank, F. C. and Kasper, J. S., 1958, *Acta Crystallogr.* **11**, 184; 1959, *Acta Crystallogr.* **12**, 483.

**Table 13.1  Some Natural Elements Which Have an "fcc" or "hcp" Crystalline Structure**

| fcc | hcp |
|-----|-----|
| Al | Be |
| Ag | Cd |
| Ar (20 K) | Co |
| Au | He (He$^4$, 2 K) |
| Ca | Gd |
| Ce | Mg |
| $\beta$-Co | Re |
| Cu | Ti |
| Ir | Zn |
| Kr (58 K) | |
| La | |
| Ne (20 K) | |
| Ni | |
| Pb | |
| Pd | |
| Pt | |
| $\delta$-Pu | |
| Rh | |
| Sr | |
| Th | |
| Xe (92 K) | |
| Yb | |

*Note:* These are at room temperature unless otherwise indicated.

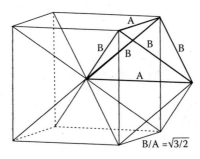

**Figure 13.10**  The bcc structure may be regarded as a packing of tetrahedra of the kind indicated here.

In the Frank–Kasper structures the packed spheres that may be placed at the vertices of tetrahedra have 12, 14, 15, or 16 neighbors, and the average for any particular Frank–Kasper structure is between 13.33... and 13.5.

Each tetrahedron of a tetrahedrally-packed structure has four other tetrahedra sharing its faces, so the dual structure, derived by placing vertices in the center of the tetrahedra (i.e., in the interstices between spheres), forms a four-connected network. Such a network can be regarded as the packing of polyhedra which are restricted to pentagonal and hexagonal faces only. These networks also represent significant structures in chemistry—the clathrates (Section 11.9).

## 13.8  Changed Utterly: Quasicrystals

We have seen in the previous paragraphs how, starting from such clues as the regular angular shapes of crystals, scientists constructed a theory—crystallography—in which the intrinsic structure of crystals was described as a periodic assembly of atoms, in an apparently complete system of ordered structures in the solid state. But in November 1984 a revolution took place: Shechtman, Blech, Gratias, and Chan announced a new state of condensed matter, found in rapidly solidified *AlMn* alloys. The new state is ordered but nonperiodic. The scientific journal *Physics Today* headlined "Puzzling Crystals Plunge Scientists into Uncertainty." Marjorie Senechal, a mathematician who has been one of the protagonists of this revolution, described this climate of astonishment in her book *Quasicrystals and Geometry*[8]:

> It was evident almost immediately after the November 1984 announcement of the discovery of crystals with icosahedral symmetry that new areas of research had been opened in mathematics as well as in solid state science. For nearly 200 years it had been axiomatic that the internal structure of a crystal was periodic, like a three-dimensional wallpaper pattern. Together with this axiom, generations of students had learned its corollary: icosahedral symmetry is incompatible with periodicity and is therefore impossible for crystals. Over the years, an elegant and far-reaching mathematical theory had been developed to interpret these "facts." But suddenly—in the words of the poet W. B. Yeats—*all is changed, changed utterly.*

What terrible beauty was born? These new solids, with diffraction patterns which exhibit symmetries that are forbidden by the crystallographic

---

[8] Senechal, M., 1995, *Quasicrystals and Geometry* (Cambridge: Cambridge University Press).

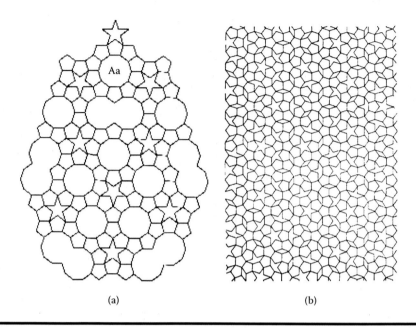

(a)                              (b)

**Figure 13.11** Two nonperiodic tilings: That proposed by Kepler in 1619 (a) and that proposed by Penrose in 1974 (b). Note that Kepler's one is finite whereas Penrose's can be continued on the whole plane.

restrictions, have been called *quasicrystals*. The internal structure of a quasicrystal is an ordered packing of identical local configurations with nonperiodic positions in space.

With this perspective, it is a shock to realize that the seed of quasicrystallinity was already there in Kepler's work.[9] In his book *Harmonices Mundi* (1619) Kepler described a repetitive structure with the "forbidden" fivefold symmetry, but with "certain irregularities."

> If you really wish to continue the pattern, certain irregularities must be admitted, (...) as it progresses this five-cornered pattern continually introduces something new. The structure is very elaborate and intricate.

Hundreds of years later, in 1974, Roger Penrose produced a tiling that can be considered the realization of the one described by Kepler (Figure 13.11).[10]

---

[9] Or even older instances in Islamic architecture: See Makovicky, E., 1992, *Fivefold Symmetry*, (ed.), I. Hargittai (Singapore: World Scientific), p. 67.

[10] Penrose, R., 1974, Pentaplexity, *Bull. Inst. Math. Appl.* **10**, 266–271.

The Penrose tiling covers the entire plane; it is nonperiodic but repetitive. *Nonperiodic* means that if one takes two identical copies of the structure there is only one way to superimpose one exactly on the other. In other words, sitting in a given position, the landscape around, up to an *infinite* distance, is unique and cannot be seen from any other point. *Repetitive* means that any local part of the structure is repeated an infinite number of times in the whole structure.

Exactly 10 years after Penrose's work, this kind of order was first observed in nature.

What most surprised the researchers after the discovery of quasicrystals was that these structures have diffraction patterns with well-defined sharp spots that were previously considered to be the signature of periodicity. How can aperiodic structures produce a pattern of sharp spots in diffraction? Let us just say that the condition to have diffraction is associated with a strong form of repetitiveness, which is typical of these quasicrystalline structures.

Mathematically, these quasiperiodic patterns can be constructed from a crystalline periodic structure in a high-dimensional hyperspace by cutting it with a plane oriented with an appropriate angle with respect to the crystalline axis. This construction explains the existence of diffraction peaks but does not offer any physical understanding.

Why has nature decided to pack atoms in these nonperiodic but repetitive quasicrystalline configurations? This is a matter of debate, unlikely to be quickly resolved.[11]

As a consequence of this revolution it was necessary for the scientific community to ask: *What is to be considered to be a crystal?*

The International Union of Crystallography established a commission on Aperiodic Crystals that, in 1992, proposed the following definition for "crystal":

> *A crystal is any solid with an essentially discrete diffraction diagram.*

According to this edict, the definition of a crystal which we have given earlier is too restrictive but the new one will, alas, be a mystery to many. One wonders how long the committee argued over that evasive word "essentially."

Penrose tilings are now to be found in many mathematics and physics departments, and elsewhere. Their use by a manufacturer of toilet paper led to a legal suit. Even mathematics is no longer immune to the intrusion of lawyers.

---

[11] See, for instance, Steinhardt, P., Jeong, H.-C., Saitoh, K., Tanaka, M., Abe, E., and Tsai, A. P., 1998, Experimental verification of the quasi-unit-cell model of quasicrystal structure, *Nature* **396**, 55–57; 1999, *Nature* **399**, 84.

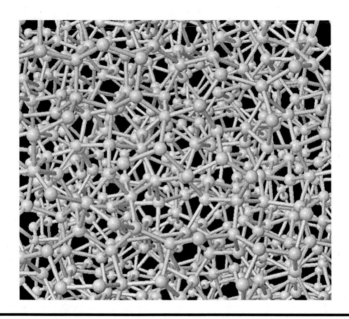

**Figure 13.12** A model for amorphous silicon. (Courtesy of Norman Mousseau.)

## 13.9 Amorphous Solids

Since ancient times, the distinction has been made between crystals and noncrystalline or amorphous (shapeless) solids (Figure 13.12). But the first category was reserved for large crystals such as gemstones. It was not realized until the present century that most inanimate materials (and quite a few biological ones as well) consist of fine crystalline grains, invisible to the eye and not easy to recognize even under a microscope. Hence, they were wrongly classified as amorphous. Despite important clues, such as the fracture surface of typical metals, crystallinity was regarded as the exception rather than the rule.[12]

It remains difficult to distinguish an aggregate of very fine crystals from an amorphous solid using x-ray diffraction, and much research has been founded on an increasingly meaningless distinction. Today, such amorphous materials as window glass are accepted as having a disordered structure, just as John Tyndall suggested with typical lyricism in *Heat—A Mode of Motion* (1863):

> To many persons here present a block of ice may seem of no more interest and beauty than a block of glass; but in reality it

---

[12] See for general reference: Lines, M. E., 1994, *On the Shoulders of Giants* (Bristol: Institute of Physics).

bears the same relation to glass that orchestral harmony does to the cries of the marketplace. The ice is music, the glass is noise; the ice is order, the glass is confusion. In the glass, molecular forces constitute an inextricably entangled skein; in ice they are woven to a symmetric texture (...)

Amorphous metals, usually obtained by very rapid cooling from the liquid state, also have a disordered structure, in this case approximated by Bernal's random sphere packings (Chapter 3). This structure gives them exceptional properties, useful in magnetic devices or as a coating on razor blades and in the manufacture of state-of-the-art golf clubs. What was once a mere academic curiosity now caresses the chin of the aspiring executive and adds several meters to his drive into the shrubbery from the first tee. Further meters will be added in Chapter 18.

## 13.10   Crystal Nonsense

Old errors cast long shadows in our conception of the natural world. Astrology still commands copious column-inches in the daily papers, spiritualists ply a busy trade, and books abound on the mystical power of crystals. A suitable crystal, we are told, can radiate its energy to us and influence our aura, in a harmonious vibration of happiness. If only it were so easy... solid state physicists would go around with permanent smiles on their faces.

This strange attribution of a hidden potency is derived from the time when crystals were regarded as rare exceptions to the general disorder of inanimate nature. Their strange perfection of form must have led primitive man to wonder at them: it is known that Peking Man collected rock crystals. Perhaps in modern times the renewed fascination with "crystal energy" also derives from the period of early radio when "crystal sets," consisting of little besides a point contact to a semiconducting crystal, acting as a rectifier, could be used to listen to radio broadcasts from strange and distant worlds. Magic indeed!

To the scientist there was a more substantial mystery in crystals. The details of the process of growth of a crystal proved intractable to any convincing explanation for some time. The problem was, for example, mentioned by A. E. H. Tutton (1924) in *The Natural History of Crystals*:

One of the most deeply interesting aspects of a crystal (...) concerns the mysterious process of its growth from a solution (...). The story of the elucidation, as far as it has yet been accomplished, of the nature of crystallization from solution in water is one of the most romantic which the whole history of science can furnish.

It is, by its very nature, a delicate problem of surface science, where minute amounts of impurities, or defects, can do strange things to help or hinder growth. But, by and large, crystallization is no longer such an enigma—otherwise the semiconductor industry could hardly be engaged in making huge silicon crystals of extraordinary perfection, everyday.

The progress of physics did not, however, deflect Rupert Sheldrake from using the assumed intractability of the explanation of crystallization as a pretext for his provocative and popular theory of "morphic resonance," according to which crystallization is guided by a memory from the past. As a botanist of impeccable pedigree, he was then able to generalize his principle to the living world, creating a considerable following and almost unbearable irritation in the orthodox scientific community. This eruption of mystical speculations caused the editor of *Nature* to jocularly remark that he would set aside his deepest principles and declare that Sheldrake's book should be burned.

The human soul seems to crave mystery at least as much as the brain demands rationality.

# Chapter 14

# Apollonius and Concrete

## 14.1  Mixing Concrete

Any builder knows that to obtain compact packings in granular mixtures such as the "aggregate" used to make concrete, the size of the particles must vary over a wide range. The reason is evident: small particles fit into the interstices of larger ones, leaving smaller interstices to be filled, and so on. A typical recipe for very dense mixtures starts with grains of a given size and mixes them with grains of smaller and smaller sizes in prescribed ratios of size and quantity. The resulting mixture has a packing fraction (fraction of volume occupied by the grains) that can approach unity.

Such recursive packing was already conceived around the 200 B.C. by Apollonius of Perga (269–190 B.C.), a mathematician of the Alexandrine school. He is classified with Euclid and Archimedes among the great mathematicians of the Greek era. His principal legacy is the theory of those curves known as conic sections (ellipse, parabola, hyperbola). He brought it to such perfection that 1800 years passed before Descartes recast it in terms of his new methods.

The method of recursive packing later reappeared in a letter by Leibniz (1646–1716) to Brosses:

> Imagine a circle; inscribe within it three other circles congruent to each other and of maximum radius; proceed similarly within each of these circles and within each interval between them, and imagine that the process continues to infinity. (Figure 14.1)

Something similar also arose in the work of the Polish mathematician, Waclaw Sierpiński (1887–1969), who wrote a paper in 1915 on what has come to be called the "Sierpiński Gasket," and it is well known as a good

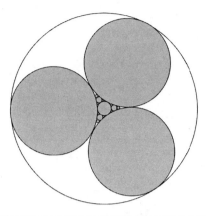

**Figure 14.1** Apollonian packing.

example of a fractal structure (see Figures 14.2 and 14.3). Ian Stewart has called it the "incarnation of recursive geometry."

The aim of Sierpiński was to provide an example of a curve that crosses itself at every point, "a curve simultaneously Cartesian and Jordanian of which every point is a point of ramification." Clearly this curve is a *fractal*, but this word was coined by Benoît Mandelbrot[1] only in 1975.

Sierpiński was exceptionally prolific—he published 720 papers and more than 60 books. He called himself an "Explorer of the Infinite."

Fractals entered in the world of physicist for the first time in the work of E. E. Fournier d'Albe writing in 1907 on the distribution of matter in the universe.[2] He is one of those minor figures of science that deserves to be better remembered—as a physicist, writer, linguist, pan-Celtic leader and inventor (including the first transmission of a picture by telegraph).

## 14.2  Apollonian Packing

In the packing procedure known as "Apollonian Packing," one starts with three mutually touching circles and inserts in the hole between them a fourth circle which touches all three. Then the same procedure is iterated.

Apollonius studied the problem of finding the circle that is tangent to three given objects (each of which may be a point, line, or circle). Euclid had already solved the two easiest cases in his *Elements*, and the other (apart from the three-circle problem) appeared in the *Tangencies* of Apollonius. The three-circle problem (or the kissing-circle problem) was finally

---

[1] Mandelbrot, B. B., 1977, *The Fractal Geometry of Nature* (New York: Freeman).
[2] Fournier d'Albe, E. E., 1907, *Two New Worlds* (Longmans: London).

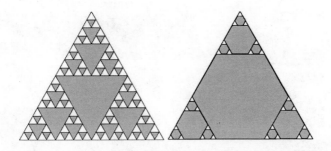

**Figure 14.2**   (Left) Sierpiński Gasket. (Right) Another fractal recursive packing made of hexagons.

solved by Viète (1540–1603) and the solutions are called Apollonian circles (Figure 14.4). A formula for finding the radius ($r_4$) of the fourth circle which touches three mutually tangent circles of radii ($r_1$, $r_2$ and $r_3$) was given by René Descartes in a letter in November 1643 to Princess Elisabeth of Bohemia:

$$2\left[\left(\frac{1}{r_1}\right)^2 + \left(\frac{1}{r_2}\right)^2 + \left(\frac{1}{r_3}\right)^2 + \left(\frac{1}{r_4}\right)^2\right]$$
$$= \left[\left(\frac{1}{r_1}\right) + \left(\frac{1}{r_2}\right) + \left(\frac{1}{r_3}\right) + \left(\frac{1}{r_4}\right)\right]^2. \tag{14.1}$$

This formula was rediscovered in 1936 by the physicist Sir Frederick Soddy who expressed it in the form of a poem, "The Kiss Precise"[3]:

> *Four pairs of lips to kiss maybe*
> *Involves no trigonometry.*
> *'Tis not so when four circles kiss*
> *Each one the other three.*
> *To bring this off the four must be*
> *As three in one or one in three.*
> *If one in three, beyond a doubt*
> *Each gets three kisses from without.*
> *If three in one, then is that one*
> *Thrice kissed internally.*
>
> *Four circles to the kissing come,*
> *The smaller are the benter,*
> *The bend is just the inverse of*
> *The distance from the centre.*

---

[3] Soddy, F., 1936, The kiss precise, *Nature* **137**, 1021.

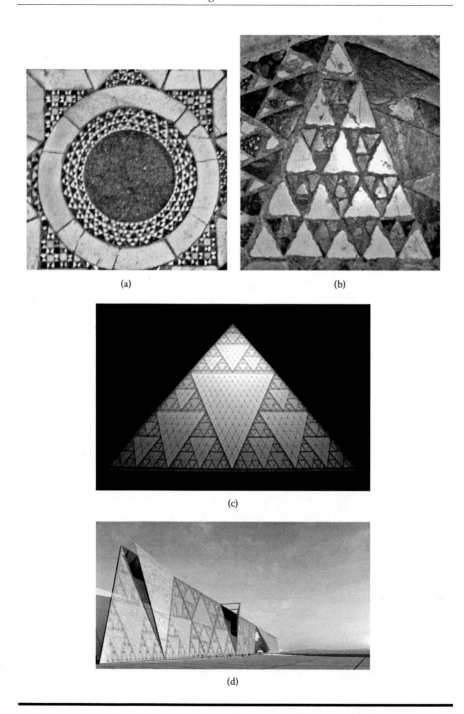

**Figure 14.3** (a,b) Fractal tessellation from Monreale Cathedral (about 1200, Sicily). (c,d) The new Grand Egyptian Museum at the site of the pyramids also incorporates the Sierpiński motif. (Courtesy of Heneghan Peng Architects.)

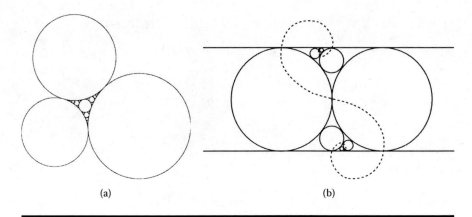

(a)                                                              (b)

**Figure 14.4** Two examples of fractal packings: The Apollonian packing (a), the loxodromic sequence of circles (b). (See Coxeter, H. S. I., 1966, Loxodromic sequences of tangent spheres, *Aeq. Math* **1**, 104–121.)

*Though their intrigue left Euclid dumb.*
*There's now no need for the rule of thumb.*
*Since zero bends a straight line*
*And concave bends have minus sign,*
The sum of the squares of all four bends
Is half the square of their sum.

*To spy out spherical affairs*
*An oscular surveyor*
*Might find the task laborious,*
*The sphere is much the gayer,*
*And now besides the pair of pairs*
*A fifth sphere in the kissing shares.*
*Yet, signs and zero as before,*
*For each to kiss the other four*
The square of the sum of all five bends
Is thrice the sum of their squares.

In the Apollonian procedure, the size of the circles inserted inside the holes become smaller and smaller and the packing fraction approaches unity in the infinite limit. For example, one can start from three equal tangent unit circles which have a packing fraction of 0.907.... Inside the hole one can insert a circle with radius $1/6.46... = 0.15...$ and the packing fraction becomes 0.95.... Now there are three holes where one can insert three circles with radius $1/15.8... = 0.063...$ and the packing fraction rises to 0.97....

## 14.3 Packing Fraction and Fractal Dimension

Pursued indefinitely, Apollonian packing leads to a dense system with packing fraction $\rho = 1$. But how is this limit reached? Suppose for instance that we start with circles of radii $r_{large}$ and stop the sequence when the radii arrive at the minimum value $r_{small}$. The packing fraction is

$$\rho = 1 - p_0 \left( \frac{r_{small}}{r_{large}} \right)^{(2-d_f)} \tag{14.2}$$

where $d_f$ is the fractal dimension and $p_0$ is the initial porosity. Indeed the Apollonian packing is a classical example of a fractal, in which the structure is composed of many similar components with sizes that scale over an infinite range. Numerical simulations give $d_f = 1.305\ldots$[4] The analytical determination of $d_f$ is a surprisingly difficult problem. Exact bounds have been calculated by Boyd[5] for the so called "Apollonian Gasket"[6] who found $1.300\,197 < d_f < 1.314\,534$. The fractal dimension can be calculated for the packing with triangles in the Sierpiński Gasket which has $d_f = 1.585$. The other packing with hexagons shown in Figure 14.2 has instead $d_f = 1$.[7]

## 14.4 Packing Fraction in Granular Aggregates

The Apollonian packing procedure can be extended to three dimensions. In this case, four spheres are closely packed touching each other and a fifth one is inserted in the hole between them. And so on.

Descartes' theorem (Equation 14.1) was extended to three dimensions by Soddy in the third verse of his poem and to $d$ dimensions by Gosset in another poem also entitled "The Kiss Precise."[8]

---

[4] Manna, S. S. and Hermann, H. J., 1991, Precise determination of the fractal dimension of Apollonian packing and space-filling bearings, *J. Phys. A: Math. Gen.* **24**, L481–L490.

[5] Boyd, D. W., 1973, Improved bounds for the disk packing constants, *Aeq. Math.* **9**, 99–106; Boyd, D. W., 1973, The residual set dimension of the Apollonian packing, *Mathematika* **20**, 170–174.

[6] Consisting in an Apollonian packing filling with Soddy circles, the hole between three, equal mutually tangent circles.

[7] Bidaux, R., Boccara, N., Sarma, G., de Seze, L., de Gennes, P. G., and Parodi, O., 1973, Statistical properties of focal conic textures in smetic liquid crystals, *J. Physique* **34**, 661–672; Eggleton, A., 1953, *Proc. Cam. Phil. Soc.* **49**, 26.

[8] Gosset, T., 1937, The kiss precise, *Nature* **139**, 62.

**Figure 14.5** "So we may imagine similar rings of spheres above and below (...) and then being all over again to fill up the remaining spaces and so on *ad infinitum*, every sphere added increasing the number that have to be added to fill it up!" (Soddy, F., 1937, The bowl of integers and the hexlet, *Nature* **139**, 77–79. With permission.)

> *And let us not confine our cares*
> *To simple circles, planes and spheres,*
> *But rise to hyper flats and bends*
> *Where kissing multiple appears.*
> *In n-ic space the kissing pairs*
> *Are hyperspheres, and Truth declares—*
> *As n + 2 such osculate*
> *Each with n + 1 fold mate.*
> The square of the sum of all the bends
> Is *n* times the sum of their squares.

Gosset's equation replaces the factor 2 in front of Equation (14.1) with a factor *d* (the space dimension, *n* in Gosset's poem). In three dimensions it gives, for example, a radius of 0.2247... for the maximum sphere inside the hole between four touching unit spheres.

The relation for the packing density can also be extended to three and higher dimensions[9]

$$\rho = 1 - p_0 \left( \frac{r_{\text{small}}}{r_{\text{large}}} \right)^{(d-d_{\text{f}})}. \tag{14.3}$$

---

[9] Herrmann, H. J., Mantica, G., and Bessis, D., 1990, Space-filling bearings, *Phys. Rev. Lett.* **65**, 3223–3226.

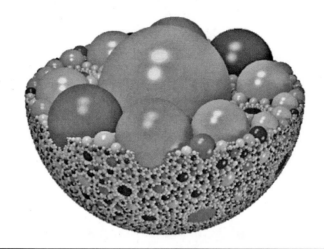

**Figure 14.6** A computer simulated Apollonian packing. (Courtesy of R. M. Baram and H. J. Herrmann.)

This expression is not limited to the Apollonian case but is valid for any fractal packing in the limit of very wide polydispersity ($r_{\text{large}} \gg r_{\text{small}}$) with power law in the size distribution.

The fractal dimension for Apollonian sphere packings has been less studied than in the two-dimensional case. The two models of packings with triangles and hexagons can be extended to three dimensions giving $d_f = 2$ and $d_f = 1.26$. Computer simulations (Figure 14.6) give an estimate $d_f = 2.4739...$[10]

Engineers have long known[11] that the porosity $p$ of the grain mixture is a function of the ratio between radii of the smallest and the largest grains utilized. They empirically use the equation

$$p = 1 - \rho = p_0 \left( \frac{r_{\text{small}}}{r_{\text{large}}} \right)^{\frac{1}{5}}. \tag{14.4}$$

A comparison with Equation (14.3) gives the fractal dimension as $d_f = 3 - \frac{1}{5} = 2.8$ for the best recipe to make strong concrete.

[10] Borkovec, M., de Paris, W., and Peikert, R., 1994, The fractal dimension of the Apollonian sphere packing, *Fractals* **2**, 521–526.

[11] Caquot, M. A., 1937, Le role des matériaux inerts dans le béton, *Société des Ingeniéurs Civils de France*, pp. 563–582.

# Chapter 15

# Packings and Kisses in High Dimensions

## 15.1 Packing in Many Dimensions

The world of mathematics is not confined to the three dimensions of the space that we inhabit. Mathematicians study sphere-packing problems in spaces of arbitrary dimension. Geometrical puzzles can be posed and solved in such spaces. Some practical challenges end up in such a form.

With this chapter we present a short excursion into the outer space of high dimensions. The topic is explored in a very comprehensive way by Conway and Sloane in their book *Sphere Packings, Lattices and Groups*,[1] which is considered by many to be the bible of this subject.

Packings in many dimensions find applications in number theory, numerical solutions of integrals, string theory, theoretical physics and digital communications. In particular, some problems in the theory of communications, with a bearing on the optimal design of codes, can be expressed as the packing of $d$-dimensional spheres. Indeed, in signal processing it is convenient to divide the whole information into uniform pieces and associate each piece with a point in a $d$-dimensional space (a point in a $d$-dimensional space is simply a string of $d$ real numbers $\{u_1, u_2, u_3, \ldots, u_d\}$). To transmit and recover the information in the presence of noise one must ensure that these points are separated by a distance larger than that at which the additional noise would corrupt the signal. Each point (a piece of encoded information) can be seen as surrounded by a finite volume, a $d$-dimensional

---

[1] Conway, J. H. and Sloane, N. J. A., 1988, *Sphere Packings, Lattices and Groups* (Berlin: Springer).

ball with a diameter larger than the additional noise. The encoded information can be reliably recovered only if these balls are nonoverlapping. An efficient coding, which minimizes the energy necessary to transmit the information, organizes these balls in the closest possible packing. Therefore, the problem relates back to the greengrocer's dilemma: *How can we arrange these balls most tightly? What is the maximum packing fraction for solid spheres in* d *dimensions?*

It might be thought that this journey into many dimensions might be uneventful: *If you've seen one such space, you've seen them all?* Surely, some simple generalization of the close-packing strategy which works in three dimensions can be successfully extended? Not so. The possibilities are much richer than this and include some startling special cases. Who would have thought that dimensions 8 and 24 are very special? In these cases structures of particularly high packing fraction can be found. The 8-dimensional one is called $E_8$ and dates from the last century, while the 24-dimensional one was discovered by J. Leech of Glasgow in 1965. New dense structures, indeed whole families of them, continue to be constructed. For example, in 1995, A Vardy announced a new packing record for 20 dimensions.

We have already seen in Chapter 3 how close-packed hexagonal layers of spheres may be stacked to generate the face-centered cubic lattice, the one described by Kepler as "the most compact solid." This is the regular packing with the highest packing fraction in three dimensions. This procedure also works in many dimensions: $d$-dimensional compact structures are often stacked one upon the other (or laminated) making an $(d + 1)$-dimensional packing. There are some special dimensions where these laminated lattices have remarkable properties in terms of packing fraction and symmetry. A very special one is $d = 24$, the laminated lattice discovered by Leech in 1965. It has a packing fraction of $\rho = 0.00193\ldots$ and it is conjectured to be the densest *lattice* packing in 24 dimensions. This packing is highly symmetrical and the symmetry group associated to it has fundamental importance in the history of group theory. Indeed, the discovery of the Leech lattice in 1968 resulted in the discovery of three new simple groups.

Group theory is an abstract branch of mathematics which is closely tied to practical applications throughout physics. Simple groups make a special finite class of groups from which any finite group can be built up. The largest simple group, the "Monster," was constructed by R. L. Griess in 1981, using the Leech lattice. The classification of all simple groups was finally completed in 1982, after having involved the work of many mathematicians for more than 50 years.

Low-dimensional sections of the Leech packing produce laminated packing in dimensions $d < 24$ but curiously they also produce nonlaminated ones. Among them is the $K_{12}$ lattice first described by Coxeter and Todd in 1954, which is likely to be the densest lattice packing in $d = 12$.

**Table 15.1    Densest Known Lattice Packings Up to Dimension 24**

| Dimension d | Packing Fraction ρ | Lattice |
|---|---|---|
| 0 | 1 | $\Lambda_0$ |
| 1 | 1 | $\Lambda_1$, Z |
| 2 | $\frac{\pi}{2\sqrt{3}} = 0.906\ldots$ | $\Lambda_2$, $A_2$ |
| 3 | $\frac{\pi}{3\sqrt{2}} = 0.740\ldots$ | $\Lambda_3$, $A_3$ |
| 4 | $\frac{\pi^2}{16} = 0.616\ldots$ | $\Lambda_4$, $D_4$ |
| 5 | $\frac{\pi^2}{15\sqrt{2}} = 0.465\ldots$ | $\Lambda_5$, $D_5$ |
| 6 | $\frac{\pi^3}{48\sqrt{3}} = 0.372\ldots$ | $\Lambda_6$, $E_6$ |
| 7 | $\frac{\pi^3}{105} = 0.295\ldots$ | $\Lambda_7$, $E_7$ |
| 8 | $\frac{\pi^4}{384} = 0.253\ldots$ | $\Lambda_8$, $E_8$ |
| 12 | $\frac{\pi^6}{19440} = 0.0494\ldots$ | $K_{12}$ |
| 16 | $\frac{1}{16}\frac{\pi^8}{8!} = 0.0147\ldots$ | $\Lambda_{16}$ |
| 24 | $\frac{\pi^{12}}{12!} = 0.001\,93\ldots$ | $\Lambda_{24}$ |

(Courtesy of J. H. Conway.)

It has been proved that the laminated lattices ($\Lambda_d$) are the densest lattice packings up to dimension $d = 8$. They are the densest known up to dimension 29 except for $d = 10, 11, 12, 13$, and it seems likely that $K_{12}$, $\Lambda_{16}$ and $\Lambda_{24}$ (the Leech lattice) are the densest in dimensions 12, 16, and 24. (See Table 15.1.)

Note that even in the "very dense" 24-dimensional Leech lattice, the volume occupied by the spheres is less than 0.2% of the total. Indeed, the packing fraction of sphere packings tends to zero as the dimension goes to infinity.

"Packings of Spheres Cannot Be Very Dense" is the title given by C. A. Rogers to a chapter of his book *Packing and Covering* (1964) where he calculates a bound for the packing fraction of sphere packings in any dimension. His bound is calculated from the generalization to high dimensions of the three-dimensional case of four spheres in mutual contact with centers on the vertices of a regular tetrahedron (Section 3.5). In $d$ dimensions this configuration is made of $d + 1$ mutually contacting spheres. In the large $d$ limit it has a packing fraction of

$$\rho \sim \frac{d}{e}\left(\frac{1}{\sqrt{2}}\right)^d. \tag{15.1}$$

These local configurations of $d+1$ spheres cannot be tightly assembled in the $d$-space. Some interstices always remain and so $\rho$ is an upper bound for the density of $d$-dimensional sphere packings.

Other more restrictive bounds have been given during the years. Summarizing these results, the packing fraction of the closest $d$-dimensional packing is between the bounds[2]

$$(2d)2^{-d} \lesssim \rho \lesssim (1.5146\ldots)^{-d} \qquad (15.2)$$

for large $d$. One can see that when the dimension increases by 1, the packing fraction of the closest packing (lattice or non lattice) is divided by a number between 2 and 1.546... and therefore goes rapidly to 0.

## Magic Dimensions

To understand why there are magic dimensions for which very special lattice packings appear, let us consider the simple case of the hypercubic lattice packing. In this packing, spheres with unit radius are placed with the centers at positions $(2u_1, 2u_2, \ldots, 2u_d)$ where $u_i$ are integer numbers. In two dimensions this is the square lattice packing, and the elementary local configuration is a set of four spheres with the centers on the vertices of a square. In $d$ dimensions the elementary local configuration of this packing is a $d$-dimensional hypercube with $2^d$ spheres on its vertices. It is easy to see that infinite copies of this local configuration form the whole packing. Each interstice between the $2^d$ spheres can accommodate a sphere with maximum radius $\sqrt{d} - 1$. In two dimensions this radius is $\sqrt{2} - 1 \sim 0.41$, in three dimensions it is $\sqrt{3} - 1 \sim 0.73$, but in four dimensions this radius is $\sqrt{4} - 1 = 1$, which means that the sphere inserted in the interstice inside the cube can have the same unit radius of the external ones. Now, two copies of the cubic packing can be fitted together without overlap to form a new lattice packing with a packing fraction that is double the original one. This is known as checkerboard lattice $D_4$, and is possibly the densest lattice in four dimensions. In this packing, the spheres have centers in the points $(0, 0, 0, 0)$, $(1, 1, 0, 0)$, $(1, 0, 1, 0) \ldots$. In general in a packing $D_d$ the spheres have centers in $(u_1, u_2, u_3, u_4, \ldots)$ where the $u_i$ are integers that add to an even number.

It is the doubling possibility that renders special the dimensions 8 and 24. Indeed, in $d = 8$ the packing $D_8$ can be doubled in two copies that fit together making the lattice packing $E_8$, the densest in $d = 8$. Analogously, in 24 dimensions it is a doubling of a particular lattice packing associated with the Golay code that forms the Leech lattice packing.

---

[2] The lower bound is by Ball, K., 1992, *Int. Math. Res. Notices* **68**, 217–222. Whereas, the upper bound is by Kabatiansky, G. A. and Levenshtein, V. I., 1978, *Problems of Information Transmission* **14**, 1–17.

## 15.2 A Kissing Competition

*How many spheres can be placed around a given sphere, such that they are all of the same size and touch the central one?* Mathematicians often refer to such contact as "kissing." This was the topic of a famous discussion between Isaac Newton and David Gregory in 1694. Newton believed that the answer was 12, as in the compact packing of Kepler, while Gregory thought that 13 was possible. The problem is not simple. The solid angle occupied by an external sphere is less than 1/13 of the total and the volume around the central sphere which is available to touching spheres is, in principle, sufficient to contain the volumes of 13 spheres.[3] But the correct answer is 12.

Note that a configuration of 13 spheres around a central one makes a very compact packing without quite achieving contact of all spheres with the central one. This is one of the local configurations that locally pack more tightly than the Kepler bound (Chapter 3).

The 12 spheres can be disposed in several positions. Examples already encountered in this book are the rhombic dodecahedron configuration, in the icosahedral arrangement (where the 12 spheres are separated from each other and touch only the central one) or in the pentahedral prism (with 2 spheres at the north and south poles and 5 spheres around each of them). The first configuration is the most compact global packing (the Kepler close packing). The other two are packings that are *locally* more dense than the first one but that cannot be continued in the whole space.

## 15.3 Kissing the Neighbors in Higher Dimensions

As Newton and Gregory did for three dimensions, one can ask which is the greatest value of the kissing number $(\tau)$ that can be attained in a $d$-dimensional packing of spheres. In one dimension the answer is clearly two, in two dimensions it is six and in three dimensions the answer is 12. The answer is unknown for dimensions above four except for $d = 8$ and $d = 24$. In 8 dimensions spheres arranged in the lattice packing $E_8$ touch 240 neighbors, and in 24 dimensions each sphere kisses 196 560 neighbors in the Leech lattice $\Lambda_{24}$. (See Table 15.2.)

Dimension $d = 9$ is the first dimension in which nonlattice packings are known to be superior to lattice packings. Here, $\Lambda_9$ has $\tau = 272$ whereas the best bound known is 380. Remember that the kissing number question aims to find the best *local* configuration (one sphere and its surroundings) while the sphere packing problem is a *global* problem.

---

[3] In fact, the fraction is $1/13.397\,33\ldots$, the same number which appears in the lower bound on the average number of cells in a foam of minimal surfaces (Chapter 11).

**Table 15.2  Known Lower and Upper Bounds for the Kissing Numbers Up to Dimension 10 and for $d = 24$**

| Dimension d | Lower Bound | Upper Bound |
|---|---|---|
| 1 | **2** | |
| 2 | **6** | |
| 3 | **12** | |
| 4 | **24** | |
| 5 | 40 | 46 |
| 6 | 72 | 82 |
| 7 | 126 | 140 |
| 8 | **240** | |
| 9 | 306 | 380 |
| 10 | 500 | 595 |
| 24 | **196,560** | |

*Note:* Boldface indicates bounds which are realized.

Bounds for the maximal attainable kissing numbers have been calculated and refined during the years. For large dimensions, we expect the kissing numbers to be between the bounds

$$2^{0.2075d} \le \tau \le 2^{0.401d[1+o(1)]} \tag{15.3}$$

(where o(1) indicates and unknown number which must have a value of *order* 1).

In Figure 15.1 the values of $\tau$ for configurations with a high kissing number are reported together with the bound from Equation (15.3) with exponent fixed at 0.802d.

## 15.4  Will Disorder Win in the End?

It is common experience that when things are well organized in an orderly fashion they occupy less space. We have seen that this is indeed also the case for sphere packings in three dimensions. Some known results, such as that for Leech lattice $\Lambda_{24}$ (Section 15.3), indicate that ordered crystalline packings are also the most compact for higher dimensions.

However, it seems that for even higher dimensions disorder might take the lead. This is what Salvatore Torquato and Frank Stillinger are currently conjecturing and backing up with some numerical demonstrations.[4] They found that the first disordered packing that is denser than known lattice

---

[4] Torquato, S. and Stillinger, F. H., 2006, New conjectural lower bounds on the optimal density of sphere packings, *Exp. Math.* **15**.

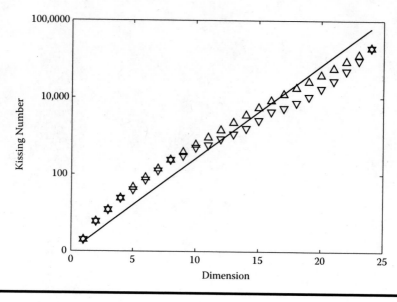

**Figure 15.1**   Known upper and lower bounds for the kissing numbers (symbols) and the Kabatiansky and Levenshtein bound $2^{0.802d}$ (line).

packings, appears at dimension $d = 56$. They have also conjectured that the maximal packing fraction can be larger than the following lower bound

$$\rho \geq 3.276\ldots d^{1/6}2^{-0.778\ldots d}, \tag{15.4}$$

which improves appreciably upon the bound in Equation (15.2). Predicting, for instance, samples more than 100,000 times denser than the known lattice packings at dimension $d = 1000$. The same conjecture also improves the prediction of the maximal kissing number to $\tau \geq 40.248\ldots d2^{0.2213\ldots d}$.

*Now order with disorder doth contend*
*to win the packing prize in higher d*
*and none can tell the outcome in the end*
*just as in Master Shakespeare's tragedie.*

## Chapter 16

# The Sweets in the Jar, the Pebbles on the Beach...

## 16.1 Those Sweets

We asked at the outset: how many sweets in the jar? Since then we have rambled inconclusively around an answer for hard sphere packings, in the usual academic evasion. It depends how you fill it, and how you shake it, and...

But in reality the sweets may not even be perfectly round, although some of us have spent happy hours reducing the diameters of spherical brandy balls and gobstoppers.

One of today's favorite candies comes in the form of an oblate spheroid, much celebrated in TV ads. These are the M&Ms™, roughly equivalent to the transatlantic British Smarties.

This is the simplest departure from sphericity that we can contemplate in a candy. (Liquorice Allsorts can wait.) How then do M&Ms pack?

One might well suppose that the extra complication of a nonspherical shape would lower the packing density. It certainly surprised many people when Paul Chaikin and his undergraduate students undertook the obvious experiment. They found a slightly higher density than that of hard spheres. How can this be?

What we might call the Chaikin Effect has been explained as follows. The nonspherical shape give the sweets an extra possibility for maneuver to increase density—they may rotate. Rotation of spheres does nothing.

This argument can be put in persuasive form, by starting with any sphere packing and then stretching all the spheres by an equal amount in one direction. The packing fraction is unchanged. Now in a second step, we rotate

**Figure 16.1** M&Ms. (From Wikimedia. With permission.)

the spheres in whatever ways will increase the packing fraction. It is not immediately clear that they can rotate, but a reexamination of the argument of constraints that was originally developed by Maxwell (Section 12.3) indicates that this should be possible.

## 16.2 Hey, What Shape Do You Want Your Ice Cubes?

Strictly, an ice cube ought to be a cube, but there is no international convention to that effect. Or, at least we think not. In practice ice cubes come in many shapes, and a certain amount of thought and post hoc rationalisation is applied to them. It is, after all, a big market.

A familiar kind is the "crescent" that comes tumbling out of machines in some hotels. This shape was chosen for reasons to do with the mechanics of production, but has been considered to have good "nestling" or packing properties. This implies a high packing fraction, and hence less gin in your martini, viewed as a desirable feature from the other side of the bar. On the other hand, "clumping" has been judged to be undesirable (presumably from our side of the bar), and a "top hat" shape is now offered, with that in mind.

Having downed that martini, it is time for some fresh air...

## 16.3 Another Walk on the Beach

On a New Year break in Cyprus, one of us (DW) strolled on the beach at the reputed birthplace of Aphrodite, Godess of Desire. In the manner of Reynolds, he admired first the sand and then the foam, but then a third element presented itself in beautifully shaped pebbles. Such stones, characteristic of shingle beaches, are roughly ellipsoidal, with three different aspect ratios, which are fairly consistent. How come?

The answer presumably lies in the continual abrasion of the stones as they grind together under the action of the waves. The original stones can be random in shape, but they have their corners polished off, since these will feel more impacts. But how does that particular beautiful shape get chosen, eventually? It must have something to do with packing and its probabilities: the stones lie together in close order, like the M&Ms, and are jostled rather than widely separated by the waves. Thus emerges a beautiful shape from the void, just like Aphrodite. At least one group is engaged in simulating this process.

Yet more philomorphic delights await us on another coast: let us hurry to Ireland to see them, as many have done before.

## Chapter 17

# The Giant's Causeway

## 17.1 Worth Seeing?

The Giant's Causeway (Figure 17.1) is a columnar basalt formation on the north coast of Ireland. It has been an object of admiration for many centuries and the subject of continual scientific debate.[1] Although it still draws tourists from afar, Dr. Johnson's acid remark that it was "worth seeing but not worth going to see," echoed by the irony of Thackeray's account of a visit, may be justified, since similar geological features occur throughout the world. Among the more notable examples are those of the Auvergne in France, Staffa in Scotland, and the Devil's Postpile in the Sierra Nevada of California. There is also the town of Stolpen ("columns") in Germany, where there sits on a hill (of basalt columns) a castle which incorporates material from the columns. It includes a dungeon, in which an incarcerated geologist could find many features of interest, to pass the time.

The primary historical importance of the debate on the origins of the Causeway lies in it being a focus of the intellectual battle between the Neptunists and the Vulcanists in the eighteenth century. The history of geology delights in giving such titles to its warring sects: others have been classed as Plutonists, Catastrophists, and Uniformitarians.

To the convinced Neptunist, the origin of rocks lay in the sedimentary processes of the sea, while a Vulcanist would argue for volcanic action. As in most good arguments, both sides were right in certain cases. But in the case of basalt, the Neptunists were seriously wrong.

---

[1] Herries Davies, G. L., 1981, *A Geology of Ireland*, (ed.), C. H. Holland (Edinburg: Scottish Academic).

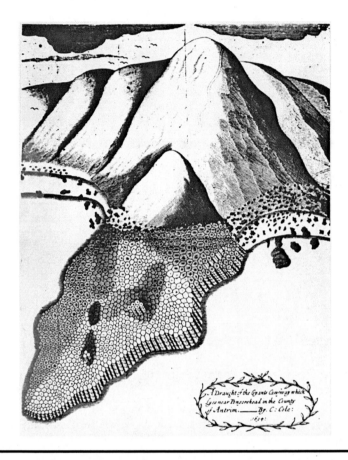

**Figure 17.1** Sketch of the Giant's Causeway. (From *Philosophical Transactions of the Royal Society,* 1694. With permission.)

Why should this concern us here? Simply because the story is intertwined with many of the strands of ideas on packing and crystallization in the understanding of materials which emerged over the same period.

*What was so fascinating about the Causeway?*

## 17.2 Idealization Oversteps Again

As with the bee's cell, but even more so, sedentary commentators on the Causeway have generally overstated the perfection of order which is to be seen in the densely packed basalt columns. They have been described as "hexagonal," implying that the pattern is a perfect honeycomb. This is not at all the case (see Figure 17.2).

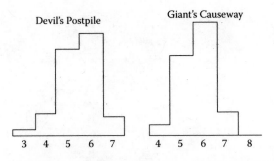

**Figure 17.2**  Distribution of the number of sides of basalt columns in two of the most famous sites where these occur. (After Spry, A., 1962, The origin of columnar jointing, particularly in basalt flans, *J. Geol. Soc. Australia* **8**, 191. With permission.)

Eyewitness reports, especially by unscientific visitors, were generally more accurate. The Percy family of Boston reported in their *Visit to Ireland* (1859):

> Do they not all look alike?

> Yes, just as the leaves are alike in general construction, but endlessly diverse, just as all human faces are alike, but all of them possessed of an individual identity.

But as the story was passed around, idealization constantly reasserted a regular hexagonal pattern for the Causeway, instead of the elegant random pattern in which less than half of the polygons are six-sided, as one can verify from Figure 17.2.

"Hexagonal" was an evocative word, calling to mind the form of many crystals. It was natural, therefore, to call the Causeway "crystalline"; and see the columns as huge crystals, even though their surfaces were rough in appearance. Alternatively, it was suggested that the columns were formed by compaction or by cracking. The latter was the choice of the Vulcanists, who saw the basalt as slowly cooling and contracting, until it cracked.

That random cracking should have this effect seems almost as unlikely as crystallization, at first glance, yet it has come to be accepted, as we explain later. This has not stopped the continued generation of wild theories, even late in the twentieth century.

## 17.3  The First Official Report

In 1693, Sir Richard Bulkeley made a report of the Causeway to the Royal Society in London. Like many who were to follow, he offered an account of the phenomenon without troubling himself to go to see it. He relayed

**Figure 17.3**  Polyhedral basalt columns in the Giant's Causeway.

the news from a scholar and traveler well known to him, that it "consists all of pillars of perpendicular cylinders, Hexagones and Pentagones, about 18 to 20 inches in diameter" (Figure 17.3).

While offering no promise to make a visit himself, he offered to answer any queries. In the following year, Samuel Foley published answers to questions forwarded by Bulkeley. Already, the similarity to crystalline forms was noted. A more scholarly and verbose account by Thomas Molyneux of Dublin followed, complete with classical references, and containing some intemperate criticisms of the original reports. Despite his superior tone, the author confesses that "I have never as yet been upon the place myself." He noted a similarity to certain fossils described by Lister but found the difference of scale difficult to explain away.

For a time the arguments lapsed, but a number of fine engravings with detailed notes were published. Art served science well, in providing an inspiring and accurate picture for the armchair theorists of geology. The correspondence resumed around 1750. Richard Pockock had spent a week at the Causeway, and settled on a Neptunist mechanism of precipitation.

In 1771, N. Desmarest published a memoir which was to be central to the Vulcanist/Neptunist dispute. In his view the "regular forms of basalt are the result of the uniform contraction undergone by the fused material as it cooled and congealed." This was countered by James Keir, who reasserted the crystalline hypothesis, drawing on observations of the recrystallization of glass. Admittedly there was a great difference of scale, but "no more than is proportionate to the difference observed between the little works of art and the magnificent operations of nature." Here, "art" means craft or industry, for recrystallization was a matter of intense commercial interest in the attempt to reproduce oriental porcelain.

The Reverend William Hamilton added further support to crystallization with published letters and a dreadful 100-page poem (*Come lonely Genius of my natal shore...*), published in 1811. This and vitriolic rebuttals of Desmarest by Kirwan and Richardson did not succeed in reversing the advance of the Vulcanist hypothesis. It was, said Richardson, an "anti-Christian and anti-monarchist conspiracy," since it set out to "impeach the chronology of Moses." He favored a model of compression of spheroidal masses to form columns, with some laboratory experiments to back it up.

It was probably fair comment when Robert Mallett summarized the state of play in 1875 by saying that "no consistent or even clearly intelligible theory of the production of columnar structure can be found."

Mallett was an early geophysicist, who invented the term "seismology." The name of his engineering works still adorns the railings of Trinity College Dublin. He undertook a thorough review of the basalt question and attempted to publish it in the *Proceedings of the Royal Society*. After 5 months and four referees he was told that "it was not deemed expedient to print it at present"—a splendidly diplomatic refusal to publish the work of a Fellow of the Society.

This reversal may well have stemmed from his trenchant criticism of "very crude and ill-thought-out notions" and a "bad or imperfect experiment inaccurately reasoned upon and falsely applied" by his predecessors, who were "blinded by a preconceived and falsely based hypothesis." His own advocacy was directed in support of contraction and cracking. He gave credit for this to James Thomson (the Glasgow professor who was the father of Lord Kelvin) together with the French school of Desmarest.

He finally succeeded in publishing his article in the lesser (and less conservative) journal *Philosophical Magazine*.

## 17.4 Mallett's Model

Mallett proposed to attack the problem "in a somewhat more determinate manner." By this he meant that his approach would be mathematical and quantitative, in contrast to the hand-waving of the other geologists. This more modern style has made the article influential ever since.

One of his principal interests was in the energy liberated in volcanic eruptions, so it was natural for him to think in terms of the total energy of the system of cracks rather than their precise mechanism of formation. At a time when there was a general tendency to express the laws of physics as minimal principles, he appealed rather vaguely to the "principle of least action" and "the minimum expenditure of work." *What crack pattern would minimize energy?*

This question makes little sense unless some constraint fixes the size of the cells of the pattern, but this was somehow ignored, and he triumphantly announced that the hexagonal pattern was best, by comparison with other simple cases.

## 17.5    A Modern View

D'Arcy Wentworth Thompson recognized that cracks which proceed explosively from isolated centres could never form such a harmonious pattern. He failed to see the possibility that crack patterns could propagate very slowly inwards as the rock cooled. A careful reading of Mallett's paper shows that he had already recognized this, and this part of his description is impeccable.

In their slow motion the cracks migrate until they form a balanced network which propagates unchanged. To have this property it need not be ordered: the best analogy is the arrangement of atoms when a liquid becomes a glass on cooling. There is local order only, reminiscent of (but not equivalent to) the local rules of equilibrium of the two-dimensional (2-D) soap froth (Chapter 10).

When precisely the realization dawned more generally that it must be so is not clear, but certainly Cyril Stanley Smith gave this explanation in 1981, when preparing the published version of a lecture to geologists. It appears very natural to the modern mind.

## 17.6    The Last Word?

Up to recently the explanation in terms of propagating cracks has been little more than appealing narrative. This changed in 2005 when Goehring and Morris[2] published an extensive account of a model system that reproduces the main features of the Causeway very well in the laboratory (Figure 17.4). It consists simply of a slurry of ordinary cornstarch, allowed to dry (and hence contract) under carefully controlled conditions, and analyzed with three-dimensional (3-D) visualization techniques.

This is an old experimental system, according to the authors: they cite Thomas Huxley's 1881 work *Physiography*. But previous use of it has been

---

[2] Goehring, L. and Morris, S. W., 2005, Order and disorder in columnar joints, *Europhys. Lett.* **69**, 739–745.

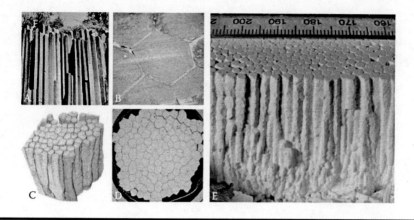

**Figure 17.4** (a) The colonnade of the Devil's Postpile, California. (b) Exposed surface of the Devil's Postpile, showing the quasihexagonal fracture pattern, as it occurs in basalt. (c) Microcomputed x-ray tomography image of cornstarch colonnade with 2.5 cm/side. (d) Cross-section of a tomogram at a depth of 18 mm. (e) A typical cornstarch colonnade (shown inverted). (Courtesy of L. Goehring and S. W. Morris.)

mainly devoted to observation of the cracking of a shallow layer of this material, which shows the familiar T-junction pattern of parched mud.

To mimic the Causeway, a deep specimen is used. As it dries it is first broken up by large cracks, then by smaller ones that propagate slowly downwards. To match the (more or less) uniform basalt colums it was found necessary to control drying in such a way that the cracks advanced with constant velocity. This is a surprise, at odds with one's ideas of heat diffusion, and may be due to the penetration of groundwater.

The laboratory measurements match the Causeway remarkably well, apart from the different scale of the starch columns (millimeters), and this is well understood. This success extends to various incidental features. The Giant's Causeway problem may be laid to rest. Computer simulations are expected to put another nail in the coffin soon.

## 17.7   Lost City?

In October 1998, an article in the British press reported the possible discovery of a lost city by a documentary filmmaker in Nicaragua. The evidence consisted of 62 polygonal basalt columns. Troops had been dispatched to guard the find against looters. Geologists had expressed some scepticism. . . .

Columnar basalt, once attributed by the Irish to a legendary giant, still astonishes and captures the imagination.

# Chapter 18

## Finite Packings and Tessellations, from Soccer to Sausages

### 18.1 The Challenge of a Finite Suitcase

Up to this point we have been concerned with packings of an infinite number of objects. The real world, and real experiments, cannot accommodate such infinities, and there are always boundary effects to be considered, in practice. Many of the experiments that we have cited involve corrections for the effects of surfaces (see Section 5.3).

Packing a small number of objects in a specific finite container poses a different kind of problem, more familiar in everyday life.

There are, as we shall see, many variations on this theme, from mathematical sausages to a popular design of soccer ball.

### 18.2 Soccer Balls

A favorite close-up of the television sports director shows a soccer ball distending the net. This offers the opportunity to compare the two: for the net is nowadays made in the form of the hexagonal honeycomb, and the surface of the ball looks roughly similar. Closer examination usually reveals the presence of 12 pentagons among the hexagons on the ball.

The problem of the soccer ball designer was to produce a convenient polyhedral form which is a good approximation to a sphere. The presently favored design replaces a traditional one and we are not aware of the precise arguments that brought this about. Certainly it is more aesthetic, on

**Figure 18.1** Icosahedral symmetry in soccer: 12 pentagonal and 20 hexagonal tiles pack on a close surface making the familiar soccer ball.

account of its high symmetry, which is officially described as *icosahedral*. The simplest design of this type would be that of the pentagonal dodecahedron (Figure 8.5), but this was perhaps not sufficiently close to a sphere. Instead, 32 faces are used, of which 12 are pentagons and the rest are hexagons (Figure 18.1).

By a curious coincidence, this icon of modern sport has cropped up in a prominent role in modern science as well, as we shall see shortly. But first let us switch sports and examine the golf ball.

## 18.3   Golf Balls

The dynamics of a sphere immersed in a viscous fluid presents one of the classic set-piece problems of physics and engineering, dating back to the work of Newton (applied, in particular, to the motion of the Earth through the ether), and tidied up in some respects by Sir George Gabriel Stokes more than 100 years ago. In modern times it can be safely assumed that there has been a large investment in a better understanding of the motion of a sphere in air, since it commands the attention of many important people on the golf courses of the world.

The extraordinary control of the golf ball's flight which is exercised by (some) golfers owes much to the special effects imparted by the spin which is imposed on the ball by the angled blade of the club. This can be as much as 10,000 revolutions per minute. Because of its backspin the ball is subject to an upward force, associated with the deflection of its turbulent wake. This is the Magnus Effect, also responsible for the swerve of a soccer ball. It causes the ball to continue to rise steeply until, both velocity and spin having diminished, it drops almost vertically onto its target.

It was found that marking the surface of the ball enhanced the effect, and a dimple pattern (Figure 18.2) evolved over many years. Today, such

**Figure 18.2**   Close-packed dimples on a golf ball.

patterns typically consist of 300 to 500 dimples. All are consistent with the incorporation of a "parting line" where two hemispherical moulds meet to impress the shape. Locally, the dimples are usually close-packed on the surface of the ball.

We know of no physical theory which would justify any particular arrangement. Many are used, whether motivated by whim or the respect for the intellectual property of established designs. Titleist has favoured an essentially icosahedral ball derived from the pentagonal dodecahedron by adding hexagons, just as for the soccer ball. Note that here we are dealing with a "dual" structure: each dimple lies at the center of one of the polygons. A simple topological rule governs all such patterns. Indeed, to wrap a honeycomb on a sphere or other closed surface is not possible, without introducing other kinds of rings. According to Euler's theorem the minimum price to be paid is 12 pentagons. One may take the pentagonal dodecahedron, beloved of the Greeks, and expand it by the addition of any number of hexagons (except, as it happens, one). In some cases the result is an elegantly symmetric structure, which will be seen later in this chapter.

## 18.4   Buckyballs

The modern science of materials has matured to the point at which progress seems barred in many directions. One cannot envisage, for example, magnetic solids which are much more powerful than the best of today's products, because they are close to very basic theoretical limits. (The scientist who makes such assertion is always in danger of following in the footsteps of the very great men who denounced the aeroplane, the spaceship and the exploitation of nuclear energy as patent impossibilities.)

Despite this sense of convergence to a state in which optimization rather than discovery is the goal, new materials continue to make dramatic entrances. Sometimes they arise because certain assortments of many elements have not previously been tried in chemical combination. There is one great exception to this trend towards combinatorial research. The sensational advance in carbon chemistry which goes by the affectionate nickname of "buckyballs" has raised more eyebrows and opened more doors than anything else of late, with the possible exception of high-temperature superconductors.

This is not a case of a single, momentous revelation: rather one of steadily increasing knowledge and decreasing incredulity over many years, from the first tentative clues to the establishment of major research programmes throughout the world. The story up to 1994 has been compellingly recounted by Jim Baggott in *Perfect Symmetry*.[1]

At its conclusion Baggott wondered which of the four central personalities of his tale would be rewarded by the Nobel Prize, which is limited to a trio. The answer came in 1996: the Royal Swedish Academy of Science awarded the Nobel Prize in Chemistry to Robert F. Curl, Harold W. Kroto, and Richard E. Smalley.

The carbon atom has long been renowned as the most versatile performer in the periodic table. It is willing to join forces with other atoms either three or four at a time. Pure carbon with fourfold bonding is diamond, whereas graphite consists of sheets with the honeycomb structure in which there is threefold bonding. The two solids have vastly different properties; one is hard, the other soft; one is transparent, the other opaque; one is horribly expensive, the other very cheap. Nothing could better illustrate the falsity of the ancient idea that all the properties of elements spring directly from the individual atoms. It matters very much how the ingredients are put together, just as in the kitchen.

For pure carbon, diamond and graphite were long known and that was supposed to be the end of the story, give or take a few other forms to be found under extremely high pressures. Yet we now recognize that graphite-like sheets can be wrapped to form a spherical molecule of 60 atoms—the buckyball—and buckyballs can be assembled to form an entirely new type of carbon crystal, with startling properties. And further possibilities continue to emerge in the laboratory or the fevered imagination of molecule designers: other larger molecules, concentric molecules like onions, tubular forms called nanotubes, which may be key components of future nanoengineering. The buckyball belongs to an infinite family of possibilities which comprise the new subject of fullerene chemistry.

The buckyball has a great future, a fascinating history and even an intriguing prehistory. The existence of molecules like this had been teasingly conjectured by a columnist in the *New Scientist*, and Buckminster

---

[1] Baggott, J., 1994, *Perfect Symmetry: The Accidental Discovery of Buckminsterfullerene* (Oxford: Oxford University Press).

**Figure 18.3** The $C_{60}$ buckyball.

Fuller built most of his reputation on the architectural applications of such structures (generally containing many more hexagons than the buckyball). Hence, the $C_{60}$ molecule was first baptized *buckminsterfullerene* in his honor. This is by no means a large mouthful by chemical standards, but the snappier "buckyball" (Figure 18.3) has steadily gained currency at its expense.[2]

## 18.5 Buckminster Fuller

Buckminster Fuller[3] has been described by an admirer as a "protean maverick." For many, he is the prime source of insight into many of the structures which we have pondered in this book. Over several decades he poured forth a torrent of ideas and assertions which combined the Greek faith in geometry as lying at the heart of all nature with vague and superficially impressive notions of energy and synergy. The resulting potpourri is inspiring or bewildering, according to taste. When he ventures into fundamental descriptions of nature, it is reminiscent of the speculations of those nineteenth-century ether theorists.

One of Fuller's assertions was that all nature is "tetrahedronally coordinated." Here, he was thinking of close-packing of spheres (although this, in part, contradicts the statement—not all arrangements are tetrahedral). He seems to have claimed to have discovered the ideal close-packed

---

[2] It has been suggested that, since the polyhedral structure of the buckyball and soccer ball date back to Archimides, "archiball" might be more appropriate.

[3] See, for reference: Marks, R. W., 1960, *The Dymaxion World of Buckminster Fuller* (New York: Reinhold Publishing Corp.); Rothman, T., 1989, *Science à la Mode, Physical Fashions and Fictions* (Princeton, NJ: Princeton University Press).

arrangement, only to find it in the work of Sir W. L. Bragg, who he then supposed to have independently found it around 1924!

It is the practical outcome of Fuller's ruminations, in the form of the geodesic dome, that we remember today. Here again, there is some question of priority. Tony Rothman, in *Science à la Mode*, has pointed out that such structures were patented by the Carl Zeiss company, for the construction of planetariums, in the 1920s. Fuller's patent is dated 1954. Rothman generously gives the protean maverick the benefit of the doubt....

## 18.6 Graphenes

The graphite form of carbon consists of flat sheets of atoms in the hexagonal honeycomb arrangement. A sheet of this kind ("graphene") can be bent and twisted and joined to itself, to form a variety of shapes and structures. A single fivefold ring of carbon bonds will give the sheet a conical form. The incorporations of more (12 in all) will enable the surface to be closed, to make the buckyball.

Alternatively, the sheet can be bent to form a cylinder, a nanotube (Figure 18.4). Such cylinders may be joined and we see an expanding vista of opportunities for nanoplumbing. We will arrive at similar cylindrical structures from a different direction in Section 18.11.

## 18.7 The Thomson Problem

In 1904, J. J. Thomson (famed for his experiments, but also a gifted theoretician) posed a mathematical problem that involved placing points on a sphere. He did so in the context of speculations about classical models of the atom, which were soon to be rendered out-of-date by quantum mechanics. But as with many other cases in this book, this obsolete physical problem has survived in its abstract mathematical form, and continues to intrigue mathematicians and challenge computer scientists. It is simply this: *What is the arrangement of N point electrical charges on a sphere, which minimizes the energy associated with their interactions?* This is just the sum of $1/r$ over all pairs of points, where $r$ is their separation. Generally speaking, the solution is neither the best packing nor the most symmetric arrangement of points. Beginning with L. Foppl, around 1910, mathematicians have accepted Thomson's challenge. Kusner and Sullivan[4] analyzed the cases $N < 20$, discovering that there is only one stable structure when $N$ is smaller than 16.

---

[4] Kusner, R. and Sullivan, J., 1997, *Geometric Topology* (ed.), W. H. Kazez (Cambridge, MA: International Press).

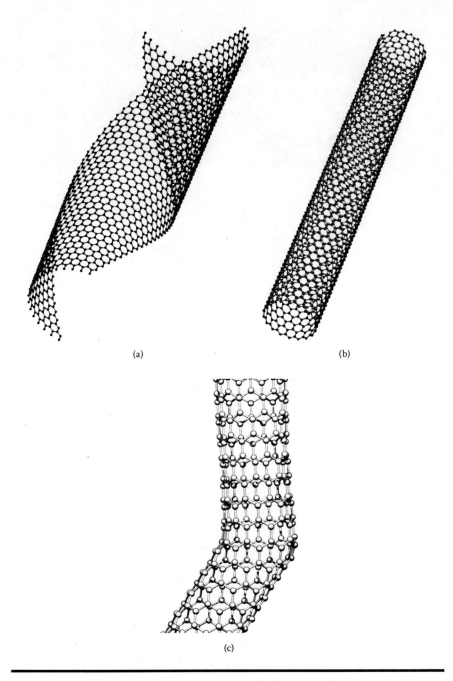

(a)

(b)

(c)

**Figure 18.4** Flat sheets of carbon atoms in the hexagonal honeycomb arrangement [graphenes (a)] can be bent to make nanotubes (b). Positive and negative curvature is induced by rings of five and seven atoms which can modify the tubular shape (c). (Courtesy of Chris Ewels, www.ewels.info.)

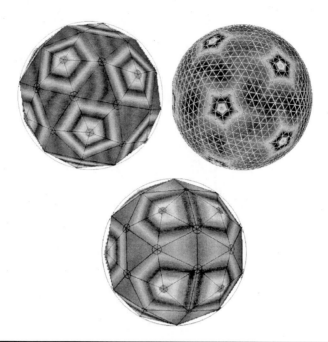

**Figure 18.5** The Thomson Problem: Three low-energy structures for charges on a sphere as depicted by Altschuler et al.[5]

The target of most researchers is to find structures of low energy using computational search procedures and identify the one which is lowest among these, for large values of $N$.[5] Some of the structures which crop up here are similar to those of Buckminster Fuller constructions and golf balls.

This computational quest is an ideal testing-ground for new software ideas, such as simulated annealing (see Appendix B).

The minimal energy structures for small $N$ are surprising in some cases: for $N = 8$ we do not find the obvious arrangement in which the charges are at the corners of a cube, but rather a twisted version of this.

For $N = 12$ the familiar icosahedral structure is found, with the charges at the corners of the pentagonal dodecahedron (Figure 8.5). Each charge has five nearest neighbors. Thereafter, this structure is adapted to accommodate more charges, for all $N$ satisfies

$$N = 10(m^2 + n^2 + mn) + 2 \tag{18.1}$$

with $m$ and $n$ being positive integers; this can be done very neatly as in Figure 18.5.

[5] Altschuler, E. L. et al., 1997, Possible global minimum lattice configurations for Thomson's problem of charges on a sphere, *Phys. Rev. Lett.* **78**, 2681.

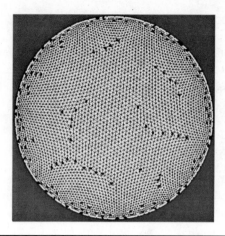

**Figure 18.6**   Packing on a disk 5000 points interacting with a Coulomb potential.

All the additional charges have six neighbors. In special cases these correspond to the buckyball or soccer ball structure which we have already admired.

That is not the end of the story. Eventually, at high $N$ these structures can be improved by modifications which introduce more local configurations of five or seven neighbors.

## 18.8   Packing Points on a Disk

The Thomson Problem is subtle because we are placing points on the curved surface of a sphere. For most forms of interaction (including the Coulomb potential in the classic version of the problem), the solution for a flat surface would be our familiar triangular lattice where each point has six nearest neighbors, but we require at least 12 defects (with five nearest neighbors) to be able to wrap it around the sphere.

Placing interacting points on a flat disk turns out to be at least as subtle. It is flat, but the curved boundary forces modifications of the triangular packing. The packing tends to be nonuniform, and thus may also disturb the perfection of triangular packing, especially when a large number of points is used. Figure 18.6 shows the result of energy minimization for 5000 points (interacting with the Coulomb potential) on a disk, as observed by Mughal and Moore.[6]

---

[6] Mughal, A., 2006, Ph.D. thesis, University of Manchester.

As well as defects at the edge there are grain boundaries in the interior. Almost certainly, the method of minimization has not quite found the true optimization in this case. In this kind of search for a minimum, the simulated annealing method (Appendix B) is used. It is often the most efficient method for finding the best minimum among many. It is a search for a needle in a haystack, or rather the best piece of straw.

## 18.9 The Tammes Problem

Many pollen grains are spheroidal and have exit points distributed on the surface. The pollen comes out from these points during fertilization. The position of the exit points is rather regular and the number of them varies from species to species. In 1930, the biologist Tammes described the number and the arrangement of the exit points in pollen grains of many species. He found that the preferred numbers are 4, 6, 8, 12, while 5 never appears. The numbers 7, 9, and 10 are quite rare and 11 is almost never found. He also found that the distance between the exit points is approximately constant, and the number of these points is proportional to the surface of the sphere.[7]

Tammes posed the following question: *Given a minimal distance between them, how many points can be put on the sphere?* We can think of the points as associated with (curved) discs of a certain size, which are not allowed to overlap. Tammes attacked the problem in an empirical way by taking a rubber sphere and drawing circles on it with a compass. He found, for instance, that when the space is enough for five circles then an extra circle can always be inserted. In this case the six circles are located at the vertices of an octahedron. In this way the preference for 4 and 6 and the aversion for 5 in pollen grains may be explained. Tammes also found that when 11 points find enough space then 12 can also be placed at the vertices of an icosahedron.

The first of these results has been mathematically proved to be valid for the surface of a sphere in three dimensions and it has been extended to any dimension. *If on the surface of a sphere in d-dimensional space more than d + 1 discs can be placed then 2d such discs can be placed at the extremities of the coordinate axes.*

The Tammes problem is the subject of an enormous amount of literature. Mathematically the questions raised by Tammes can be expressed as follows: *what is the largest diameter $a_N$ of N equal circles that can be placed on the surface of a unit sphere without overlap? How must the circles be arranged, and is there a unique arrangement?*

---

[7] Tammes, P. M. L., 1930, On the origin of number and arrangement of the places of exit on pollen grains, *Recuil d. trav. bot. néerlandais* **27**, 1.

**Table 18.1    Known Solutions of the Tammes Problem $N \leq 12$ and $N = 24$**

| N | $a_N$ | Arrangement |
|---|---|---|
| 2 | 180° | Opposite ends of a diameter |
| 3 | 120° | Equilateral triangle in the equator plane |
| 4 | 109° 28′ | Regular tetrahedron |
| 5 | 90° | Regular octahedron less one point, not unique configuration |
| 6 | 90° | Regular octahedron |
| 7 | 77°52′ | Unique configuration |
| 8 | 74° 52′ | Square antiprism |
| 9 | 70°32′ | Unique configuration |
| 10 | 66°9′ | Unique configuration |
| 11 | 63°26′ | Icosahedron less one point, not unique configuration |
| 12 | 63°26′ | Icosahedron |
| 24 | 43°41′ | Snub cube |

*Source:* H. T. Croft.[8]

Exact solutions are known only for $N \leq 12$ and $N = 24$. These and other solutions are shown in Table 18.1.

Note that for $N = 6$ and 12, $a_N = a_{N-1}$ which is the Tammes empirical result.

A bound for the minimum distance $\xi$ between any pair of points on the surface of the unit sphere, was given in 1943 by Fejes Tóth

$$\xi \leq \sqrt{4 - \csc^2 \left[ \frac{\pi N}{6(N-2)} \right]} \qquad (18.2)$$

with the limit exact for $N = 3, 4, 6$, and 12.[9]

## 18.10    Universal Optimal Configurations

Both the packing problem to which we dedicated most of this book and the Thomson problem discussed in Section 18.7 can be viewed as potential energy minimization problems. The Thomson problem minimizes the Coulomb potential whereas the packing problem minimizes some attractive potential which induce spheres to stay as close as possible but with a very steep short range repulsion that precludes overlaps. The question

---

[8] Croft, H. T., Falconer, K. J., and Guy, R. K., 1991, *Unsolved Problems in Geometry* (New York: Springer), p. 108.

[9] Ogilvy, C. S., 1994, *Excursions in Mathematics* (New York: Dover), p. 99.

presently asked by some mathematicians is whether we can find packings that are optimal for entire classes of possible potentials.

For instance, for the generalized Thomson problem it has been proved that in three dimensions there are only six possible configurations that are optimal for for *any* kind of attractive potential:

> n = 1 a single point;
> n = 2 two antipodal points;
> n = 3 an equilateral triangle on the equator;
> n = 4 a regular tetrahdron;
> n = 6 a regular octahdron;
> n = 12 a regular icosahedron.

The search for universal configurations that solve the generalized packing problem has been not successful so far, but several upper bounds, very close to realised packings, have been found. It has been conjectured that the hexagonal lattice in two dimensions; the $E_8$ and the Leech lattice are optimal in the Euclidean space.[10]

## 18.11   Helical Packings

The packing of spheres around a cylinder results in helical patterns, which may often be seen in street festivals when balloons are used to decorate lamposts. These attractive structures have an interesting history because they are found in many plants. Botanists have long been fascinated by the way in which branches or leaves are disposed along a stem, or petals in a flower. They have found many different helical arrangements, but almost all have a strange mathematical property, which is the main preoccupation of much of the extensive literature on this subject, at times acquiring a mystical flavor. This dates back (at least) to Leonardo. In the nineteenth century, the Bravais brothers, and later Airy and Tait began a more modern study, which surprisingly continues today. The reviewer of a recent book by Roger Jean said that it "remains one of the most striking phenomena of biology."

It is celebrated in an imposing structure at the heart of the Eden Project in Cornwall. In biological cases the particular spiral structures tend to involve Fibonacci numbers,[11] and this has been the subject of much speculation.

Recently, the subject has recurred in another exciting new context—the creation of nanotubes (Figure 18.4), first cousins to the "buckyballs" (see

---

[10] Cohn, H. and Elkies, N., 2003, New upper bounds on sphere packings, *Ann. Math.* **157**, 689.

[11] Numbers generated from a sequence defined recursively by: $f(0) = 0$, $f(1) = 1$, $f(n) = f(n-1) + f(n-2)$; yielding to 0, 1, 1, 2, 3, 5, 8, 13, 21, 34, 55, 89, . . . .

Section 18.4). They were first made in the 1970s by Morinobu Endo, a Ph.D. student at the University of Orleans.

Biologists call this subject "phyllotaxis," and obscure it further with terms such as "parastichies," which are rather repellent to the first-time reader. But these helical structures are really quite simple things. If we are to place spheres or points on the surface of a cylinder, it is much the same as placing them on a plane. We therefore expect to find the close-packed triangular arrangement of Chapter 2, which is optimal under a variety of conditions. The difference lies in the fact that the surface is wrapped around the cylinder and joins with itself. It is like wallpapering a large pillar—the packing must continue smoothly around the cylinder, without interruption. How we can do this with a given roll of wallpaper?

We could think of cutting out a strip from the triangular planar packing and wrapping it around the cylinder. We might have to make some adjustments to avoid a bad fit where the edges come together. We can displace the two edges with respect to each other and/or uniformly deform the original pattern (not easy with wallpaper!), in order to have a good fit.

Let us reverse this train of thought: roll the cylinder across a plane, "printing" its surface pattern again and again. We expect to get the close-packed pattern, or a strained version of it. The three directions of this pattern, in which the points line up, correspond to helices on the cylinder (Figure 18.7). Taking one such direction, how many helices do you need to complete the pattern? This defines an integer, and the three integers $k$, $l$, $m$ corresponding to the three directions can be used to distinguish this cylindrical pattern from all others. It is easily seen that two of these must add to give the third integer.

This is the notation of phyllotaxis, applied to plants, to nanotubes and to bubble packings within cylinders (Figure 18.7). In the last case, the surface structure is always of this close-packed form.

## 18.12   Virus Architecture

Structures similar to those that emerge in the Thomson Problem crop up in a corner of biology that is at the forefront of research: the structure of viruses.

Viruses come in many forms (including that of tobacco mosaic virus, which is like the spiral packing). Here we speak of that, which are roughly spherical and composed of identical units (Figure 18.8).

The modeling of these structures is closely akin to that of the Thomson Problem. In both cases, new features emerge from very large number of units. These are "scars," defective lines which reduce the cost of pasting a triangular lattice onto a sphere (Figure 18.6). We might return to our analogy with the challenge of wallpapering a curved surface, which we used for

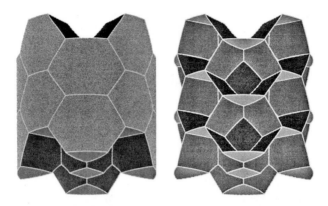

**Figure 18.7** Bubbles packed in a cylinder show the familiar hexagonal honeycomb wrapped on a cylinder. This simulation by G. Bradley shows surface patterns and the interior for one of these structures.

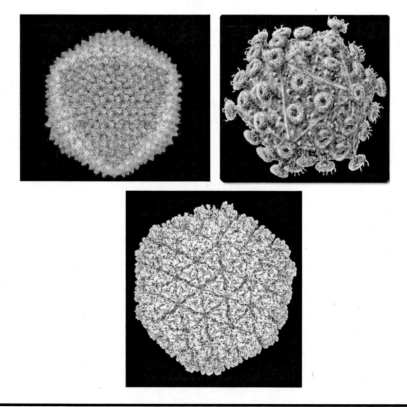

**Figure 18.8** An x-ray imaging of a virus and two models for the HIV and rice dwarf viruses (from left to right). (From Finnish Centre of Excellence in Virus Research. With permission.)

cylinders in Section 18.11. The satisfactory wallpapering of a sphere is more challenging. We can cut 12 pentagonal pieces of paper and stick them on a sphere. These pentagons are flat surfaces and nobody would be surprised to see wrinkles appearing, the equivalent of "scars." If the wallpaper is replaced by a coarse net, perhaps we could avoid the wrinkles. This corresponds to the case in which a small number of points is used, and the scars do not appear.

# 18.13 Stuffing Sausages

All packings in the real world are finite, even atoms in crystals or sand at the beach. Finite packing problems have boundaries. This makes their solution more difficult than for infinite packings.

### Optimal Box for Discs

How may we arrange $N$ unit discs so as to minimize the area of the smallest convex figure containing all discs (the convex hull)? It has been shown for all $N \leq 120$ and for $N = 3k^2 + 3k + 1$, that the convex hull tends to be as hexagonal as possible.[12]

### The Sausage Conjecture

The analogous problem in three dimensions is: *How to arrange N unit spheres so as to minimize the volume of the smallest convex figure containing all spheres?* For $N \leq 56$ the best arrangements are conjectured to be "sausages" (the centers of the spheres all along a straight line [Figure 18.9]). But larger $N$ convex hulls with minimum volume tend to be associated with more rounded clusters. This might be called the sausage–haggis transition.

For dimension $d = 4$ it has been shown that the "sausage" is the best solution for $N$ up to at least 377 000.[13] The Sausage Conjecture states that for $d \geq 5$ the arrangement of hyperspheres with a minimal volume convex hull is always a "sausage."[14]

---

[12] Wegner, G., 1986, Über endliche Kreispackungen in der Ebene, *Stud. Sci. Math. Hungar.*, 1–28.

[13] Gandini, P. M. and Zucco, A., 1992, On the sausage catastrophe in 4-space, *Mathematika* **39**, 274–278.

[14] See Croft, H. T., Falconer, K. J., and Guy, R. K., 1991, *Unsolved Problems in Geometry* (New York: Springer); Wills, J. M., 1998, Spheres and sausages, crystals and catastrophes—And a joint packing theory, *Math. Intell.* **20**, 16–21.

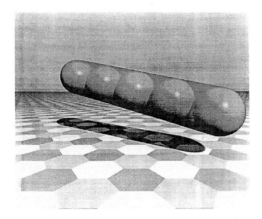

**Figure 18.9** Sausage packing of five balls. (Courtesy of J. M. Wills, University of Siegen, Germany.)

## 18.14 Filling Boxes

When objects are packed in spaces of finite size and of given shape, new questions arise. How many objects can be put inside a given box? Or equivalently, how big must a box be to contain a given set of objects? Which is the best arrangement and what is the density of this packing? Is the solution unique? These questions are all associated with the *maximum separation problem* which asks how to spread *n* points inside a given box so that the minimum distance is as large as possible.

Let us consider the two-dimensional case first.

### Discs in a Circular Box

Which is the smallest diameter $a$ of a circle which contains $N$ packed discs of diameter 1? The answer to this question is known up to $N = 10$ and conjectures are given for $N$ up to 19. For $N = 1$ the solution is clearly $a = 1$. For $2 \leq N \leq 6$ the answer is $a = 1 + 1/\sin(\pi/N)$, and it is $a = 1 + 1/\sin(\pi/(N-1))$ for $7 \leq N \leq 9$. Whereas, for $N = 10$ one finds $a = 7.747\ldots$.

### Discs in a Square

An analogous problem is to find the smallest size of the square containing $N$ packed unit circles. Exact results are known for $N \leq 9$ and $N = 14, 16, 25, 36$. Conjectures have been made for $N \leq 27$.

**Table 18.2**  $a = 1 + 1/d_N$ Where is the $d_N$ Minimal Separation between the $N$ Points in a Unit Square

| N | a (side of the square) |
|---|---|
| 1 | 1 |
| 2 | $1 + \dfrac{1}{\sqrt{2}}$ |
| 3 | $1 + \dfrac{1}{(\sqrt{2}(\sqrt{3}-1))}$ |
| 4 | 2 |
| 5 | $1 + \sqrt{2}$ |
| 6 | $1 + \dfrac{6}{\sqrt{13}}$ |
| 7 | $1 + \dfrac{1}{2(2-\sqrt{3})}$ |
| 8 | $1 + \dfrac{\sqrt{2}}{\sqrt{3}-1}$ |
| 9 | 3 |
| 14 | $1 + \dfrac{3}{\sqrt{6}-\sqrt{2}}$ |
| 16 | 4 |
| 25 | 5 |
| 36 | 6 |

*Source:* H. T. Croft.[17]

For large $N$ one has the approximate expression $a \simeq 1 + \frac{1}{2}12^{1/4}N^{1/2}$. In such a packing the density is, therefore,

$$\rho = \frac{\pi N}{4a^2} \simeq \frac{\pi}{\sqrt{12}} - \frac{\pi}{3\sqrt{\sqrt{12}N}} \tag{18.3}$$

which tends to the density of the hexagonal lattice for large $N$.

It turns out that for small $N$ the hexagonal arrangement is not the best. Square packing is better adapted to the square symmetry of the container and it results in a denser packing certainly for $N \leq 36$ and probably up to $N = 49$.[15]

Beer distributors should look into hexagonal packings now that they sell cases of 18 or more cans: the superiority of square packing is not clear for *rectangular* boxes.[16] (See Table 18.2.)

---

[15] See, for general reference: Melisson, H., 1997, *Packing and Covering with Circles* (Proefschrift Universiteit Utrecht, Met lit. opy.).

[16] Eighteen discs can be hexagonally packed in a pattern made of five lines of 4, 3, 4, 3, 4 inside a rectangular box.

[17] Croft, H. T., Falconer, K. J. and Guy, R. K., 1991, *Unsolved Problems in Geometry*

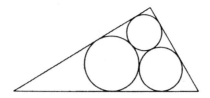

**Figure 18.10**  Malfatti's solution to the problem.

## *Spheres in Spheres*

How many spheres with unit diameters can be packed inside a given sphere? As in many other cases that we encountered in this book, very little is known. Only one sphere can fit within a sphere of diameter $\sigma < 2$. No more than two spheres can fit inside a sphere with $\sigma = 2$ and no more than 13 spheres find space within a sphere with $\sigma = 3$ (see kissing problem in Section 15.2). For larger diameters only lower bounds are known. So far, 32 spheres has been packed inside a sphere with $\sigma = 4$ and 64 spheres have been arranged within a sphere of diameter $\sigma = 5$. Other bounds are known up to $\sigma = 40$.[18] When the spherical container become very large, the number should approach the Kepler's packing which correspond to a number of inner spheres smaller than or equal to $\pi\sigma^3/\sqrt{18}$.

## 18.15  The Malfatti Problem

A parsimonious sculptor wants to cut three cylindrical columns from a piece of marble which has the shape of a right triangular prism. How should he cut it in order to waste the least possible amount of marble? The problem is equivalent to that of inscribing three circles in a triangle so that the sum of their areas is maximized.

In 1802, Gian Francesco Malfatti (1731–1807) gave a solution to this problem which thereafter bore his name. Previously, Jacques Bernoulli had given a solution for a special case. In due course, other great mathematicians were attracted to it, including Steiner and Clebsch. Malfatti assumed that the three circles must be mutually tangent and each tangent to only two sides of the triangle. Under this assumption the Malfatti solution follows as in Figure 18.10.

The problem was considered solved and for more than 100 years no-

(New York: Springer).

[18] Gensane, T., 2003, Dense packings of equal spheres in a large sphere, *Les Cahiers du LMPA J. Liuville* **188**.

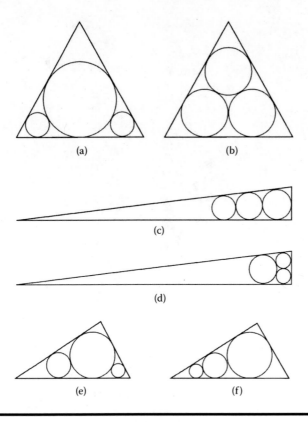

(a)  (b)

(c)

(d)

(e)  (f)

**Figure 18.11**  Solutions to the Malfatti problem.

body noticed that the Malfatti arrangement shown in Figure 18.10 is not the best. For instance, for an equilateral triangle, the solution of Figure 18.11a is better than Malfatti's one shown in Figure 18.11b. Howard Eves (1965) observed that if the triangle is elongated, three circles in line (as in Figure 18.11c) have a much greater area than those of Figure 18.11d.

Finally in 1967, Michael Goldberg showed that the Malfatti configuration is *never* the solution, whatever the shape of the triangle! The arrangements in Figure 18.11e,f are always better. Goldberg arrived at this conclusion by using graphs and calculations. A full mathematical proof has yet to be produced.[19]

---

[19] From Stanley Ogilvy, C., 1969, *Excursion in Geometry* (New York: Dover), p. 145; 1932, *Periodico di Mathematiche* **12** (4th series) is a complete review.

# Chapter 19

# Odds and Ends

## 19.1 Ordered Loose Packings

Ordered *loose* packings sometimes occur in the study of crystal structures. Here our question may be turned on its head: *What is the* lowest possible *density for a packing of hard spheres which is still mechanically stable or "rigid?"* For such rigidity, each sphere needs at least four contacts and these cannot be all in the same hemisphere. Many loose crystalline packings have been proposed. For instance, the structure shown in Figure 19.1a has $\rho = 0.1235$. It was proposed many years ago by Heesch and Laves and has been long considered to be the least dense stable sphere packing. Recently, a packing with $\rho = 0.1033$ (Figure 19.1b) was obtained by decorating, with tetrahedra, the vertices of the sodalite ($Na_4 Al_3 (Si O_4)_3 Cl$) net.[1]

The lowest known density for such stable packing is $\rho = 0.0555\ldots$[2] This value is about 10 times smaller than the one for the loose random packing. But this is not surprising, since structures such as those in Figure 19.1 are highly symmetric and cannot be obtained or approximated by simply mixing spheres at random.

## 19.2 Parking Cars

That parking can pose problems should come as little surprise to most of us.

---

[1] O'Keeffe, M. and Hyde, G. B., 1996, *Crystal Structures* (Washington, DC: Mineralogical Society of America).

[2] Gardner, M., 1966, Packing spheres, in *Martin Gardner's New Mathematical Diversions from Scientific American* (New York: Simon & Schuster).

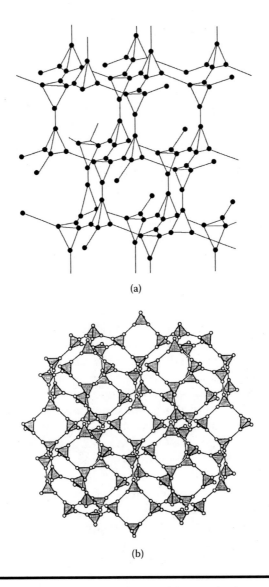

**Figure 19.1** (a) A structure (called D4) of stable equal-sphere packing with low density (the circles indicate the centers of the spheres and the lines connect neighbors in contact). (b) This structure (called W*4) is obtained by replacing the vertices of the sodalite net with tetrahedra.

For example, imagine a parking space of length $x$ where cars of unit length are parked one by one, completely at random. What, on average, is the maximum number of cars $M(x)$ that can find a place in this space? Obviously, cars are not allowed to overlap.

It is always possible to obtain a packing fraction of one by systematically putting each segment (car) in contact with two others. But if the segments are disposed in a random way without overlapping and readjustment, then after a certain stage the line has no remaining spaces large enough to accommodate another segment.

When cars are parked at random, Rényi[3] determined an integro-functional equation that gives $M(x)$ in an implicit form. In the limit of an infinitely large parking space the resulting packing fraction is

$$\rho \to 0.7475\dots \tag{19.1}$$

The packing fraction $\rho$ and the average of the maximum number of cars are simply related according to $M(x) = \rho x$.

If the random packing is modified so that a car arriving can move slightly (up to half segment length) in order to create an available space, then the packing fraction increases to 0.809.

In the analogous two-dimensional problem, objects with a unit square shape are placed in a rectangular parking lot. In this case no exact results are available. Palasti (1960) conjectured that the two-dimensional packing fraction has the same value $\rho = 0.7475\dots$ as in the one-dimensional case. But this conjecture remains unproved and computer simulations suggest that the packing fraction is slightly higher than the conjectured value.

## 19.3 Goldberg Variations

The sphere minimizes the surface area for a fixed volume, as the soap bubble teaches us. In three-dimensional packing problems we need to consider shapes which fit together to fill space, and the problem of minimizing surface area is not so easy. One interesting clue to the best strategy was provided by Michael Goldberg in 1934.

Goldberg restricted himself to the case of a single polyhedron (not necessarily space filling) with $N$ planar sides. What kind of polyhedron is best, in terms of area?

He conjectured that the solution always has threefold (or trivalent) vertices. Bearing in mind the ideal of a sphere, it is attractive to conjecture that the solution is always a regular polyhedron, that is, one with identical faces. This is not always possible. Goldberg's conjecture, supported by a good deal of evidence, states that the solution is always at least close to being regular, in the sense of having only faces with $n$ and $n + 1$ edges.

---

[3] Rényi, A., 1958, On a one-dimensional problem concerning random space-filling, *Publ. Math. Inst. Hung. Acad. Sci.* **3**, 109–127.

**Figure 19.2** Goldberg polyhedra for $N = 12, 14, 15$, and $16$. They are the building-blocks of the Weaire–Phelan and other foam structures (Chapter 7).

In particular, the solutions for $N = 12$ and $14$ are as shown in Figure 19.2. Although there is no rigorous chain of logic making a connection with the Weaire–Phelan structure (Chapter 11), it turns out to be the combination of these two Goldberg polyhedra.

## 19.4   Packing Regular Pentagons

Pentagons cannot be packed together on the plane without leaving some free space. What is the maximum packing fraction that can be achieved?

There are several obvious ways of arranging the pentagons in a periodic structure. Figure 19.3 shows two of the densest ones. The structure with packing fraction 0.92 is thought to be the densest, and it has been found in the air-table experiments of the Rennes group (Section 2.2).

## 19.5   Dodecahedral Packing and Curved Spaces

Consider a (not necessarily ordered) packing of equal spheres. Construct around any sphere the Voronoï cell (Chapter 7). It was conjectured and proved very recently by Hales and McLaughlin that the volume of any

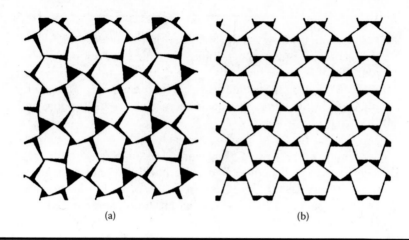

(a)                                          (b)

**Figure 19.3** Two dense packings of pentagons, with packing fractions 0.864 65 (a) and 0.921 31 (b).

Voronoï cell around any sphere is at least as large as a regular dodecahedron with the sphere inscribed. This provides the following bound for the densest local sphere packing

$$\rho \leq \rho_{\text{dodecahedron}} = \frac{V_{\text{sphere}}}{V_{\text{dodecahedron}}} = \frac{\pi \sqrt{2(29 + 13\sqrt{5})}}{5^{5/4}} = 0.7547 \ldots . \quad (19.2)$$

This is 1% denser than the Kepler packing but this is a local arrangement of 13 spheres that cannot be extended to the whole space. Indeed, regular dodecahedra cannot be packed in ordinary space without gaps, as we have seen before.

The situation is similar to that for pentagons in two dimensions. Regular pentagonal tiles cannot cover a floor without leaving any interstitial space. However, in two dimensions one can immediately see that this close packing can be achieved by curving the surface. The result is a finite set of 12 closely packed pentagons that tile the surface of a sphere making a dodecahedron. Analogously, in three dimensions, regular dodecahedra can closely pack only in a positively curved space. In this case regular dodecahedra pack without gaps making a closed structure of 120 cells which is a four-dimensional *polytope* (that is, a *polyhedron* in high dimensions).[4]

---

[4] See, for general reference: Sadoc, J. F. and Mosseri, R., 1997, *Frustration Géométrique* (Paris: Éditions Eyrolles).

It started with liquids, you know. *They* didn't understand liquids. Local geometry is non-space-filling. Icosahedra. Trigonal bipyramids. Oh, this shape and that shape, lots of them. More than the thirty-two that fill ordinary space, let me tell you. That's why things are liquid, trying to pack themselves in flat space, and that's what I told *them*.

*They* couldn't deal with it. *They* wanted order, predictability, regularity. Silly. Local geometry can be packed, I said, just not in flat space. So, I said, give them a space of constant curvature and they'll pack. All *they* did was laugh. I took some liquids to a space of constant negative curvature to show *them* it would crystallize, and it sucked me up. [Tepper, S. S., *Mavin Manyshaped* (New York: Ace Fantasy Books).]

## 19.6 Microspheres and Opals

Oranges do not spontaneously form close-packed ordered structures but atoms sometimes do. Where is the borderline between the static world of the oranges and the restless, dynamic one of the atoms, continually shuffled around by thermal energy—between the church congregation and the night-club crowd?

At or near room temperature, only objects of size less than about 1 $\mu$m are effective in exploring alternatives, and perhaps finding the best. A modern industry is rapidly growing around the technology of making structures just below this borderline, in the world of the "mesoscopic" between the microscopic and the macroscopic.

In one such line of research, spheres of diameter less than a micrometre are produced in large quantities and uniform size using the reaction chemistry of silica or polymers. Such spheres, when placed in suspension in a liquid, may take many weeks to settle as a sediment. When they do so, a crystal structure is formed—none other than the fcc packing of earlier chapters. This crystallinity is revealed by striking optical effects, similar to those which have been long admired in natural opal.

Natural opals are made of silica spheres of few hundred nanometres (1 nm = $10^{-9}$ m) in size, packed closely in an fcc crystalline array. They are valued as gemstones because their bright colors change with the angle of view. This iridescence is due to the interference of light which is scattered by the ordered planes of silica spheres. Indeed, the size of these spheres is typically in the range of visible light wavelength (430 to 690 nm).

An important goal of present research in material science is the creation of artificial structures with such a periodicity, in order to tailor their optical

**Figure 19.4** Electronic microscope image of a synthetic opal.

properties. Several studies begun in the late 1980s showed that a transparent material can become opaque at certain frequencies provided that a strong and periodic modulation of the refractive index is imposed in space. Structures with these properties have been constructed for microwave radiation but, until recently, not for visible light. With conventional microelectronic techniques it is very difficult to shape structures below 1000 nm (which is 1 $\mu$m). Artificial opals (Figure 19.4) provide the right modulation in the diffraction index, opening the way to the construction of new "photonic band gap" materials. Photonic crystals are the ingredients for future

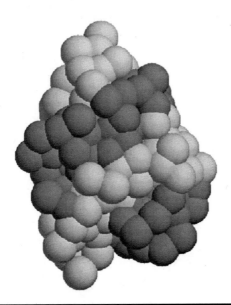

**Figure 19.5** Simulation of a multiblock copolymer formed of repeating AB units, globule phase. (Courtesy of D. Parsons and D. Williams.)

optical transistors, switches and amplifiers, promising to become as important to the development of optical devices as semiconductors have been to electronics.

## 19.7   Protein Folding

Nature routinely achieves a particular task of compaction which has intrigued and baffled physicists for decades. Proteins, which are long stringy molecules, assembled (in very many ways) from a wide variety of units, fold up when compressed by chemical forces, and sometimes assisted by "chaperone molecules," to make a compact mass. Its exact shape is essential to its eventual function, just as the shape of a key is made to fit a particular lock.

Misfoldings do occur, but not often—they have unfortunate medical consequences, such as Alzheimer's Disease.

To appreciate the problem, take a long piece of string, or—better still—a long and complicated necklace. Gather it into a ball. Obviously, if you do it again and again you will get a different result every time, as regards for the parts that are on the outside of the ball. So how does a protein find its unique appropriate form? And how does it do it in a few tens of microseconds?

Once more we may turn to the computer for a simulation which might suggest an answer. This is becoming possible today, by running the latest supercomputers for months or years. While you read this, Blue Gene may well be grinding away at this folding problem, watching the collapse of a long protein chain. The prospect of eventual success is debatable for large proteins: the folding may depend critically on the interatomic forces that are used. This is not a purely geometric problem.

Computational success may not bring much immediate insight in answering the question: What is it about the folding pathway that speeds it on its way so efficiently to its happy conclusion, instead of wandering up blind alleys or falling into traps?

Biologists mutter "evolution," as they usually do. Nature selects proteins that fold quickly, and that is that.

But physicists will continue to ponder this special attribute of the proteins that are the very stuff of life.

## 19.8   The Tetra Pak Story

A bus taking a party of physicists (which included several philomorphs and one of the present authors) to the airport after a conference in Norway in 1999 made a brief "comfort stop" at a motorway service station which included a small shop. Several of the party disappeared for a time. They had discovered something remarkable in the shop—a close packing of identical tetrahedral cartons! This clever arrangement makes sense only because it uses slightly flexible cartons and the container (which has an elegant hexagonal shape) is finite. The bus left for the airport before the contents of the container could be dissected

Tetrahedral cartons are becoming much less common and we cannot report another sighting of such hexagonal containers, but a chance encounter with a carton engineer enlightened us somewhat. It seems the reluctance of tetrahedra to pack snugly in any obvious way was indeed a recognized problem for the industry, and common practice was to pack them loosely. The intelligent design came not from the computer search of a physicist in search of a problem, but rather from an employee of the carton-making company who played with the problem and found a solution by trial and

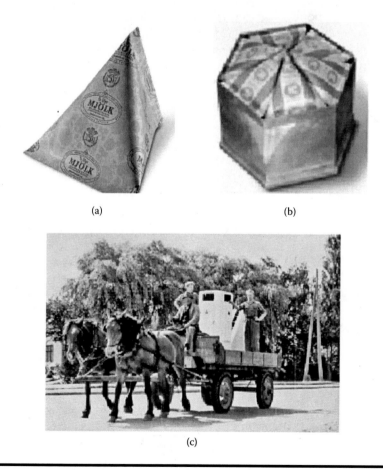

(a)

(b)

(c)

**Figure 19.6**    (a) The Tetra Pak carton, (b) the packaging, and (c) the delivery of the first Tetra Pak machine in 1952.

error. We suspect that it relates to the tcp structures of Chapter 13, but await another sighting in the field to confirm this detail.

The Tetra Pak carton itself was the invention of Ruben Rausing. As sole owner of a packaging company in Sweden in the 1950s, he had an itch to do "something that nobody else has done before." Wisely, he studied the market: What was needed? In those days milk and cream were delivered door to door in heavy milk bottles. This was a charming but inefficient daily ritual, for which some still harbor a fondness. It was surely doomed. Why not use a light-sealed carton, which could be easily picked up in the corner store? Filling these proved to be a problem, but Rausing's wife found the solution, over lunch. She pointed out that if a continuous tube was filled with milk, it could be sealed at intervals by pressure and heat from the outside. It worked. If successive sealings are at right angles, a tetrahedral pack is formed. In 1946, a prototype machine had been rigged up, but it took many years to perfect the carton material. The first production equipment was delivered to a dairy in 1952 (Figure 19.7).

Nowadays, the company mostly makes cartons with a more mundane geometrical shape, mere "bricks" of rectangular form. The original tetrahedral design still survives, however, in some limited markets, perhaps university departments of mathematics.

## 19.9 Packing Regular Tetrahedra

The packing of regular pentagons (Section 19.4), which cannot fill all space, raises an obvious question that has not been answered, to our knowledge. What is the best possible packing?

In 1970, Hoylman[5] proved that the best lattice packing with tetrahedra has density $\rho = 18/4900 = 0.36...$. The best crystalline packing so far proposed has density $\rho = 0.716...$ and it is generated by subdividing into tetrahedra a close packing of icosahedra. The tcp structures described in Section 13.7 may provide some clues. Each provides a space-filling arrangement of asymmetric tetrahedra. Within each of these we could place the largest regular tetrahedron that can be accommodated by all, obtaining a reasonably efficient packing. This could be further relaxed to obtain a slightly higher density. This strategy was recently tried by Torquato and Conway at Princeton University.[6] They discovered that the A15 structure (which was adapted to become the Weaire–Phelan structure of Chapter 11) results in a surprising low density, occupying less than 50% of the volume. Other tcp structures with some readjustments give better results but all stay below 72%.

---

[5] Hoylman, D. J., 1970, *Bull. Am. Math. Soc.* **76**, 135–137.

[6] Conway, J. H. and Torquato, S., 2006, *PNAS* **103**, 10612–10617.

While we were writing about these progresses, an announcement by Paul Chaikin at the March meeting of the American Physical Society suddenly changed our perspective. Paul Chaikin tackled the problem with the same hands-on approach that he used for the packing of ellipsoids (Section 16.1). He went to a toy shop and bought a few hundreds of tetrahedral dice and started playing with them in different containers of various shapes and sizes. He found that the dice pack with densities better than 0.75. Computer simulations and other experiments are now under investigation to confirm and support this finding but the fact that disordered packings perform better than all known ordered arrangements is surprising and it is challenging many established beliefs. It is remarkable that the known densest packing might be found by literally throwing dice.

Will disorder win in the end?

## 19.10   Nature and Geometry

It is amusing to consider what an ancient Greek, returning to our civilisation to see how the Olympic Games have developed, and picking up this book, might think of it. Would it seem as familiar as those wrestlers and foot racers? The predominance of geometry would be comfortingly familiar. Little of calculus, and none of the obscure profundities of modern topology, has intruded here. Form has been more pervasive than force, though we have tried to appreciate their relation, as form is the expression of force. As well as all those familiar polyhedra, buckyballs would delight our Greek, as would the grand Egyptian Museum, also perhaps the Water Cube (but it is not a cube, he will say), phyllotactic spirals, and much else. The Bird's Nest and other highly random creations should dismay him. But the notion that mathematics can be a source of pleasure, is akin to art, and may be embodied in it, will seem entirely natural, and eternal.

> *Music and poesy use to quicken you;*
> *The mathematics and the metaphysics,*
> *Fall to them as you find your stomach serves you.*
> *No profit grows where is no pleasure ta'en;*
> *In brief, sir, study what you most affect.*

William Shakespeare
"Taming of the Shrew"

# Appendix A

# The Best Packing in Two Dimensions

## A.1 When Equal Shares Are Best

There is a general principle which helps with many packing problems: it says that, under certain circumstances, *equal shares* are best. (We shall resist any temptation to draw moral lessons for politicians at this point.) A parable will serve to illustrate this principle in action.

A farmer, attracted by certain European subsidies, seeks to purchase *two* fields with a total area of 40 hectares. The prices for fields depend on the field size as shown in the diagram of Figure A.1. What is he to do?

Playing with the different possibilities quickly convinces him that he should buy two fields of 20 hectares each. The combination of 10 and 30 hectares is more expensive; its price is twice that indicated by the open circle on the diagram, which lies above the price of a 20 hectare field.

What property of the price structure forces him to choose equal-sized fields? It is the *upward curvature* of Figure A.1, which we may call *convexity*. We will use this property in finding the best packing of coins.

### Enclosing a Single Disc

A packing problem may be posed for a single disc of radius $R$ by requiring it to fit into a polygonal boundary—a sheep-pen—which is as small as possible. The sheep-pen has $n$ sides: what shape should it be?

The answer is: A regular polygon (that is, totally symmetric with equal edges and equal angles between edges).

To demonstrate this we note that the area of a polygon can be divided

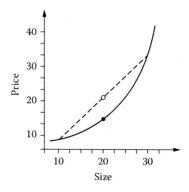

**Figure A.1**   What is the best price for two fields of a total area of 40 hectares?

into sectors, as shown in Figure A.2a. Each sector corresponds to an angle $\theta$ and the angles must add up to $2\pi$. We need to share this total among $n$ angles. But the area of each sector must be greater or equal than that shown in Figure A.2b (isosceles triangle) and this is a *convex* function of $\theta$.

So equal angles are best, and the strategy of equal shares results in $n$ such isosceles triangles.

## *Packing Many Discs*

We are ready to complete the proof of the disc-packing problem in two dimensions: *What is the most dense arrangement of equal discs, infinite in number?* The answer will be the triangular close packing of Figure 2.1. First, we assign to each disc in any given packing its own territory, a polygonal shape which surrounds it (as in Figure A.3). This is done by drawing a line bisecting the one which joins the centers of neighboring discs. This is the *Voronoï construction* described in Chapter 7.

We can assume that all these boundaries meet at a triple junction. (If not, just introduce an extra national boundary of zero length to make it so.) For an infinite number of circles in the whole plane, a theorem of Euler[1] states that for such a Voronoï pattern the average number of sides of the polygons is *exactly six*[2]

$$\langle n \rangle = 6. \tag{A.1}$$

[1] See Smith, C. S., 1981, *A Search for Structure* (Cambridge, MA: MIT Press), p. 5.
[2] There are certain kinds of tilings where this equation does not hold. See Grünbaum, B. and Rollet, G. C., 1986, *Tilings and Patterns* (New York: Freeman).

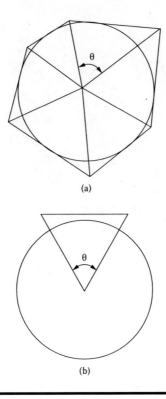

(a)

(b)

**Figure A.2** (a) A polygon with a disc inside can be divided into triangular sectors with angles $\theta$. (b) The isosceles triangle that touches the disc is the sector that minimizes the area for a given $\theta$. The area of such a sector is a convex function of $\theta$, therefore a division in equal sectors is best.

**Figure A.3** A Voronoï partition around the centers of a disordered assembly of discs.

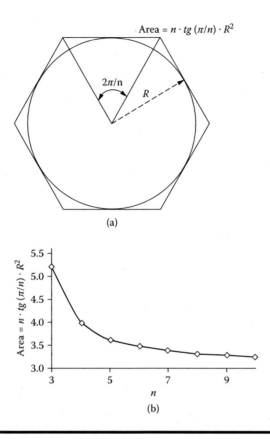

(a)

(b)

**Figure A.4** The area of regular polygons with a circle inscribed is a *convex* function of the number of sides of the polygons (*n*).

The triangular pattern has $n = 6$ for all the polygons: they are regular hexagons. If we introduce some polygons with *more* sides than six, they must be compensated by others with *less*. Now we can use the result of the last section. The area of the *n*-sided polygons cannot be less than that of the regular polygon which touches the disc. This is a convex function of $n$ (Figure A.4), so that once again we can use the principle of "equal shares," this time of polygon sides. *The total area cannot be less than what we can obtain with $n = 6$ for all polygons.*

# Appendix B

# Turning Down the Heat: Simulated Annealing

If a suit is to be made from a roll of cloth, how should we cut out the pieces in such a way as to minimize wastage? This is a packing problem, for we must come up with a design that squeezes all the required shapes into the minimum area.

No doubt tailors have had traditional rules-of-thumb for this but today's automated clothing industry looks for something better. Can a computer supply a good design?

This type of problem, that of optimization, is tailor-made for today's powerful computers. The software which searches for a solution does so by a combination of continual small adjustments towards the desired goal, and occasional random shuffling of components in a spirit of trial and error—much the way that we might use our own intelligence by a blend of direct and lateral thinking.

A particularly simple strategy was suggested in 1983 by Scott Kirkpatrick and his colleagues at IBM. Of course, IBM researchers are little concerned with the cutting (or even the wearing) of suits, but they do care passionately about semiconductor chip design, where a tiny competitive advantage is well worth a day's computing.[1] The components in microchips and circuit boards should be packed as tightly as possible, and there are further requirements and desired features which complicate the design process.

The research in question came from the background of solid state physics. Nature solves large optimization problems all the time, in

---

[1] Kirkpatrick, S., Gelatt, C. D., and Vecchi, M. P., 1983, Optimization by simulated annealling *Science* **220**, 671.

particular when crystals grown as a liquid are slowly frozen. Why not think of the components of the suit or the chip as "atoms," free to bounce around and change places according to the same spirit of laws which govern the physical world, at high temperatures? Then, gradually cool this imaginary system down and let it seek an optimum arrangement according to whatever property (perhaps just packing fraction) is to be maximized.

This is the method of "simulated annealing", the second word being taken from the processing of semiconductors, which are often heated and gradually cooled, to achieve perfection in their crystal structure.

The idea was simple and it is relatively straightforward to apply—so much so that incredulous circuit designers were not easily persuaded to try it. Eventually, it was found to be very effective, and hence it has taken its place among the optimizer's standard tools. Its skillful use depends on defining a good "annealing schedule," according to which the temperature is lowered.

The mathematical background to these large and complicated optimization problems is itself large and complicated. Complexity theory often offers the gloomy advice that the optimum solution cannot be found in any practical amount of time, by any program. But, unlike the pure mathematician, the industrial designer wants only a very good packing, not necessarily the best. Beyond a certain point, further search is pointless, in terms of profit. As Ogden Nash said: "A good rule of thumb, too clever is dumb."

# Index